The Demand Driven Adaptive Enterprise

Carol Ptak and Chad Smith

INDUSTRIAL PRESS, INC.

Industrial Press, Inc.

32 Haviland Street, Suite 3
South Norwalk, Connecticut 06854
Phone: 203-956-5593
Toll-Free in USA: 888-528-7852
Fax: 203-354-9391
Email: info@industrialpress.com

Author: Carol Ptak and Chad Smith
Title: The Demand Driven Adaptive Enterprise
Library of Congress Control Number: 2018959515

ISBN (print):   978-0-8311-3635-2
ISBN (ePUB):   978-0-8311-9495-6
ISBN (eMOBI):   978-0-8311-9496-3
ISBN (ePDF):   978-0-8311-9494-9

Editorial Director: Judy Bass
Copy Editor: Janice Gold
Compositor: Patricia Wallenburg, TypeWriting
Cover Designer: Janet Romano-Murray

Limits of Liability and Disclaimer of Warranty

industrialpress.com
ebooks.industrialpress.com

1 2 3 4 5 6 7 8 9 10

Contents

Introduction

Where does common sense turn into common *nonsense* in organizations? The basic fundamental principles we will explore in this book are supported by mathematics, economics, physics, and managerial accounting. These concepts are simply undeniable, and people readily agree on and identify them as common sense and obvious facts. Yet when we look at how organizations are actually operating, we see these principles missing or pushed to the background only to emerge through lip service or crisis. Why?

These principles make common sense, yet the application of and adherence to them are anything but common. If people can understand these ideas and agree that they make sense, what is missing? We believe we have the answer: an effective framework.

Today, organizations lack an effective framework to consistently apply and integrate common sense principles at all levels of the organization (strategic, tactical, and operational). Instead, many people inside organizations must actively fight against and/or work around their current framework just to do what they know is right. Company personnel are stressed as they are constantly forced into a no-win situation of choosing between what makes them look good or what is actually the right thing to do for the company. Worse yet, our conventional systems obscure or distort relevant information that makes knowing what is right for the company that much more difficult to determine.

We have reached a point where we will have to choose. Either your organization will continue to try to compete by doing the same old things better and faster or it will make a fundamental break from convention and try something new. The choice will be forced upon you one way or the other.

If you are ready to explore something new, this book is a good place to start. In this book we will reveal a new framework by which to run an organization in the volatile, uncertain, complex, and ambiguous (VUCA) world we live in today. This new framework is called the Demand Driven Adaptive Enterprise (DDAE) Model. Not only will we describe this model but we will detail the path and stages required to fully implement it.

This framework and book will not be embraced by everyone. It challenges a lot of conventional practices and systems. It has and will threaten those with interests in keeping those practices and systems in place. For those who have embraced this model, the results have been well worth it. So, is your organization ready for something new?

Definitions in This Book

This book will use two sources of definitions. All known and accepted terms that are not new with the advent of Demand Driven Adaptive Enterprise (DDAE) model will be defined using definitions from the fourteenth edition of the *APICS Dictionary*. The authors thank APICS for its support of this project. Since 1957, APICS has been the premier professional association for supply chain and operations management and the leading provider of research, education, and certification programs that elevate supply chain excellence, innovation, and resilience.

For terms that are new with the advent of the DDAE model, the authors have created a dictionary specific to DDMRP. Translated versions of this dictionary in multiple languages can be found at http://www.demanddriveninstitute.com.

About the Authors

Chad Smith

Chad Smith is the coauthor (with Carol Ptak) of the third edition of *Orlicky's Material Requirements Planning* (McGraw-Hill, 2011), *Demand Driven Material Requirements Planning* (Industrial Press, 2016) and *Precisely Wrong—Why Conventional Planning Fails and How to Fix It* (Industrial Press, 2017). He is also the coauthor (with Debra Smith) of *Demand Driven Performance: Using Smart Metrics*
(McGraw-Hill, 2014). He is a cofounder of and partner in the Demand Driven Institute, an organization dedicated to proliferating demand driven methods throughout the world.

In 1997 Mr. Smith cofounded Constraints Management Group (CMG), a services and technology company specializing in demand driven manufacturing, materials, and project management systems for midrange and large manufacturers. He served as Managing Partner of CMG from 1998 to 2015. Clients (past and present) include Unilever, LeTourneau Technologies, Boeing, Intel, Erickson Air-Crane, Siemens, IBM, The Charles Machine Works (Ditch Witch), and Oregon Freeze Dry. Mr. Smith is also a certified expert in all disciplines of the Theory of Constraints, studying directly under the tutelage of the late Dr. Eli Goldratt.

Chad Smith makes his home in Wenatchee, Washington, with his wife, Sarah, and two daughters, Sophia and Lily.

Carol Ptak

Carol Ptak is currently a partner with the Demand Driven Institute (www.demanddriveninstitute. com) and was most recently at Pacific Lutheran University as Visiting Professor and Distinguished Executive in Residence. Previously, she was vice president and global industry executive for manufacturing and distribution industries at PeopleSoft, where she developed the concept of demand driven manufacturing (DDM). Ms. Ptak spent four years at IBM Corporation, culminating in the position of global SMB segment executive

A leading authority in the use of ERP and supply chain tools to drive improved bottom line performance, Ms. Ptak's expertise is well grounded in four decades of practical experience as a successful practitioner, consultant, and educator in manufacturing operations. Her pragmatic approach to complex issues and dynamic presentation style has her in high demand worldwide on the subject of how to leverage these tools and achieve sustainable success.

She holds an MBA from Rochester Institute of Technology and completed the EMPO program at Stanford University. Ms. Ptak is a frequent educator at the university level and presents at many key technical conferences around the world, in places such as South Africa, France, Israel, Australia, Ireland, and the Netherlands, as well as thirteen APICS International conferences. She is the author of numerous articles and the books *DDMRP* (Industrial Press, 2016), *Orlicky's Material Requirements Planning 3e* (McGraw-Hill, 2011) with Chad Smith, in addition to *MRP and Beyond* and *ERP, Tools, Techniques and Applications for Integrating the Supply Chain, Theory H.O.W.* with Harold Cavallaro, and *Necessary but not Sufficient* with Dr. Eli Goldratt and Eli Schragenheim. Together with Dean Gilliam she updated *Quantum Leap*, originally written by John Constanza. Ms. Ptak has lent her name to the internationally coveted Ptak Prize for Supply Chain Excellence, which is awarded annually by ISCEA (International Supply Chain Education Alliance).

Ms. Ptak is certified through APICS at the fellow level (CFPIM) and was certified in Integrated Resource Management (CIRM) with the first group internationally. Ms. Ptak was the President and CEO of APICS, The Educational Society for Resource Management for the year 2000. Prior to her election as APICS President, she served on the Society in a variety of positions.

Carol Ptak currently makes her home on a working cattle ranch in the Pinal Mountains near Globe, Arizona, with her husband, Jim, her two dogs, three horses, and the largest fold of purebred registered Scottish Highland cattle in Arizona.

About the
Demand Driven Institute

With affiliates, compliant software alli- ances, and instructors throughout the world, we are changing the way busi- nesses plan, operate, think, and evolve. Your business has a choice: continue to  operate with rules, metrics, and tools developed more than fifty years ago, or make a break from convention, recognize the complex supply chains we live in, and make a fundamental change in the way it does business . . . but don't take too long or the choice will be made for you.

Thought Leadership

The Demand Driven Institute (DDI) was founded in 2011 by Carol Ptak and Chad Smith. Collectively, Ms. Ptak and Mr. Smith have authored or co-authored several published works on Demand Driven Principles, Finance and Information, and Planning Systems.

Powerful Educational Programs

DDI educational products are a powerful suite of enterprise education designed to enable companies to begin and sustain an implementation of a Demand Driven Operating Model (DDOM) and the Demand Driven Adaptive Enterprise (DDAE) model.

High Impact Simulations and Games

The Demand Driven Institute offers a suite of co-branded and fully endorsed simulations and games that teach various aspects of the Demand Driven Adaptive Enterprise Model. Simulations are offered throughout the world as both public and in-house events.

Professional Endorsement Certificates

As the leading authority on Demand Driven methods, the Demand Driven Institute offers a comprehensive series of professional endorsement certificates. These endorsement certificates are the gold standard in ensuring and identifying an individual's understanding of, and ability to apply, analyze, evaluate, and create value using Demand Driven methods.

Software Compliance

As the leading authority on Demand Driven methods, Demand Driven Institute evaluates and certifies software ensuring that a specific software has enough features and/or functions to implement, sustain and even improve a DDMRP implementation. This objective evaluation and certification is provided free of charge to software entities. If you are considering a software claiming DDMRP functionality, look for the DDI compliance label.

Learn more at www.demanddriveninstitute.com.

Acknowledgments

This book is truly built on the shoulders of many giants, from the original work of the practitioners who developed MRP, including Joe Orlicky, George Plossl, Richard (Dick) Ling, and Ollie Wight, to the great thinkers behind lean, Six Sigma, and Theory of Constraints—Taiichi Ohno, W. Edwards Deming, and Eliyahu Goldratt. The authors have stood on the shoulders of these giants to unite these different theories and methodologies and take a leap forward—into a future of planning with relevant visibility that mitigates the volatile, uncertain, and variable world that seems impossible to plan. We have known many of these giants personally and wish to express our continued appreciation to them.

Collectively, the authors would like to thank the members of the Demand Driven Institute Global Affiliate Network for a great partnership in bringing demand driven concepts to the mainstream throughout the world.

Additionally, the authors would like to thank various members of the APICS community for their amazing input and support in trying to restore the promise of effectiveness planning and information systems. Those people include Keith Launchbury, Roberta McPhail, John Melbye, Ken Titmuss, Olivia Reary, and Abe Eshkenazi.

The authors would like to point out particular individuals who have made a lasting contribution to the demand driven body of knowledge and awareness. These people include Greg Cass, Debra Smith, Erik Bush, David Poveda, Dick Ling, Paddy Ramaiyengar, Kirk Black,

Caroline Mondon, Alfonso Navarro, Dr. Patrick Rigoni, Dr. Steven Melnyk, Christoph Lenhartz, Laurent Vigouroux, Dr. Romain Miclo, and Alfredo Angrisani.

Chad Smith would like to thank his wife, Sarah, and two daughters, Sophia and Lily, for putting up with the process of writing books and courseware. The support and love of these three people has kept him going. Chad would like to thank the team at Constraints Management Group, LLC, for an amazing journey for nearly 20 years. Specifically, Chad would like to acknowledge the inspiration and accomplishments of his mother, Debra Smith; her direction-setting vision has been instrumental in articulating the Demand Driven Adaptive Enterprise model. Finally, Chad would like to thank his partner and coauthor Carol Ptak for an extremely rewarding and fulfilling partnership.

Carol Ptak would like to thank her husband, Jim, for the understanding and the continued support to keep going through the tough times. Words are so insufficient to acknowledge her parents, Dorothy and Bud, who taught her from the youngest age that she was limited only by her imagination even at a time when the glass ceiling was more like concrete. Their love and encouragement has been the wind under her wings. Carol would especially like to thank Chad Smith for an incredible experience and continued partnership—far beyond any that could ever have been imagined. Chad has opened all our eyes to the deeper truth of a new world of planning. The process of writing three books together has been an incredible journey and truly has been an honor and the highlight of a very long career.

CHAPTER 1

Don't Be a Dodo

"It is not the most intellectual of the species that survives; it is not the strongest that survives; but the species that survives is the one that is able best to adapt and adjust to the changing environment in which it finds itself."

The above quote is often falsely attributed to Charles Darwin. While undoubtedly inspired by Darwin's work, it was Leon Megginson, Professor Emeritus at Louisiana State University, who is the source of this quote. Megginson wrote several books on small business management, published over 100 articles, and won numerous awards for teaching and research.[1] Regardless of the source of the quote, the message for business leadership should be powerful: adapt or die.

The dodo bird is an extinct flightless bird that was native to the island of Mauritius, first recorded by Dutch sailors in 1598. When the Dutch colonized the island, they brought with them dogs and pigs. This resulted in an immediate and profound change to the dodo's native environment, one to which it could simply not adapt. The last credible sighting of a dodo was in 1662. In less than 100 years the dodo was gone. It disappeared so quickly many thought it was a mythical creature until researchers in the middle of the nineteenth century thoroughly studied remains of the bird.[2]

What can be learned from the dodo? The dodo had no say on the changes to the environment; they were imposed upon it. In today's world of volatility, uncertainty, complexity, and ambiguity (VUCA) there is a high degree of probability that organizations will have environmental changes imposed that will profoundly or dramatically affect their ability to compete and/or survive. This means that organizations must find a way to quickly sense and adapt to changes in the environment. What stands in the way?

Today's conventional management practices have tremendous amounts of inertia driven by software, consulting, accounting, and academic experts. Many of these practices trace their origins back to the 1930s and 1950s. Yet the world looks nothing today like it did at that time. Companies must adapt and innovate or their very existence is threatened. Consider this astonishing research from the *Harvard Business Review* in an article titled, "The Biology of Corporate Survival:"

> "We investigated the longevity of more than 30,000 public firms in the United States over a 50-year span. The results are stark: businesses are disappearing faster than ever before. Public companies have a one in three chance of being delisted in the next five years, whether because of bankruptcy, liquidation, M&A, or other causes. That's six times the delisting rate of companies 40 years ago. Although we may perceive corporations as enduring institutions, they now die, on average, at a younger age than their employees. And the rise in mortality applies regardless of size, age, or sector. Neither scale nor experience guards against an early demise.
>
> We believe that companies are dying younger because they are failing to adapt to the growing complexity of their environment. Many misread the environment, select the wrong approach to strategy, or fail to support a viable approach with the right behaviors and capabilities."[3]

But what to change to? How to change and drive adaptation? Is there a safe and effective path to transform a company from a basic planning and operational model developed in the 1950s and measured by financial accounting principles developed in the 1970s and '80s to an agile and adaptive enterprise capable of staying ahead of today's hypercompetitive markets and highly complex supply chains? This has been the focus of the authors and their organization, The Demand Driven Institute, since 2011—to articulate a comprehensive methodology that enables a company to sense changes from the market, adapt planning, production and distribution, and drive innovation in real time, resulting in sustainable and dramatic improvements to ROI.

First, the fundamentals. There are three basic necessities that management must always be carefully considering and managing in order to avoid organizational collapse and to sustain and drive better performance:

- **Working capital** is the capital of a business that is used in its day-to-day trading operations. It is typically calculated as the current assets minus the current liabilities. Important considerations include inventory levels, available cash, accounts receivable, the level of available credit, and accounts payable. It is an effective way to measure the immediate overall company health.
- **Organizational contribution margin** is the rate at which the company generates cash within certain periods. It is total revenue minus variable costs and period operating expenses.
- **Customer base** is the base of business that provides the sales volume of the organization. This includes market share, sales volume, product and/or services innovation, service levels, and quality.

Figure 1-1 depicts these three critical considerations in a strategic target chart.[4] The figure has concentric circles with a green middle, followed by a yellow ring, then a red ring, and finally a dark red ring. The area is equally divided into three sections: working capital, contribution margin, and customer base. The outer circles are the biggest cause of concern. As the measure moves farther away from the center, the threat to the organization grows as performance pushes closer to and through the "edge of collapse." The very outer ring is system collapse or failure. Any one of these three crucial necessities pushed too far outward will cause an organization to fail.

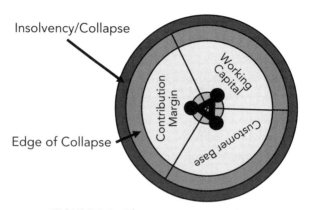

FIGURE 1-1 The strategic target chart

The black dots in Figure 1-1 represent the organization's position for each critical consideration. Conceptually, these dots will never touch. If they are consistently green, the organization and its expectations grow; the circle (and its rings) simply expands outward. Additionally, there are lines connecting the black dots depicting the tension and connections between the three areas. For example, if the customer base erodes, contribution margin and working capital will be adversely affected. If contribution margin erodes, working capital will be adversely affected.

When any one of these considerations moves into the edge of collapse ring, signal strength will intensify, and the organization will call, "All hands on deck!" to deal with that specific threat. However, there is a strong connection and tension among the three. Organizations must be careful not to overcompensate in any one area in a manner or for a duration that might drive another over the edge instead. For example, to recover acceptable customer base, the decision could be made to dramatically reduce price then affecting the contribution margin. Management must constantly fight this battle in the current highly complex and volatile environment, now and in the future—that is their primary job!

A non-business analogy is the human body, a highly complex system that must be able to perform three basic functions. The human body must be able to perform respiration (draw breath and pass oxygen to the blood) at a sufficient rate. The body then must also be able to circulate oxygenated blood throughout the body in a constant loop (pulse and blood pressure). Finally, the body must be able to maintain a fairly tight control zone of temperature or risk vital organ failure. These three basic functions essentially define what is known as the "vital signs."

The green zone centers for each of these vitals is well known throughout the medical community and will depend on certain patient characteristics such as age and sex. These green zones are defined below by the Cleveland Clinic.[5]

- **Respiratory rate.** A person's respiratory rate is the number of breaths you take per minute. The normal respiration rate for an adult at rest is 12 to 20 breaths per minute. A respiration rate under 12 or over 25 breaths per minute while resting is considered abnormal.
- **Pulse.** Your pulse is the number of times your heart beats per minute. Pulse rates vary from person to person. Your pulse is lower when you are at rest and increases when you exercise (because more oxygen-rich blood is needed by the body when you exercise). A normal pulse rate for a healthy adult at rest ranges from 60 to 80 beats per minute.
- **Blood pressure.** Blood pressure is the measurement of the pressure or force of blood against the walls of your arteries. Healthy blood pressure for an adult, relaxed at rest, is considered to be a reading less than 120/80 mm Hg. A systolic pressure of 120–139 or a diastolic pressure of 80-89 is considered "prehypertension" and should be closely monitored. Hypertension (high blood pressure) is considered to be a reading of 140/90 mm Hg or higher.

- **Body Temperature.** The average body temperature is 98.6 degrees Fahrenheit, but normal temperature for a healthy person can range between 97.8 to 99.1 degrees Fahrenheit or slightly higher. Any temperature that is higher than a person's average body temperature is considered a fever. A drop in body temperature below 95 degrees Fahrenheit is defined as hypothermia.

Any person that exhibits increasing difficulty with any one or a combination of these vital functions will be at an increasing risk of expiring. If that person was at a hospital during that difficulty there would be an escalation of monitoring, attention, and resources devoted to their care as they progress through admittance, a medical unit, a Definitive Observation Unit (DOU), and finally an Intensive Care Unit (ICU).

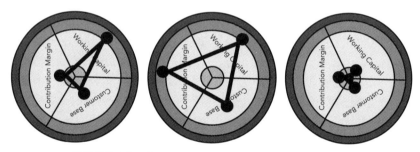

FIGURE 1-2   Alternate strategic target chart scenario

Figure 1-2 depicts three different scenarios. In each case, the relative positions of each of the three considerations are plotted. It should be pointed out that all three of these scenarios are simply a point in time reference; it could be the same company just at different points in time or it could be three different companies at one point or even different points in time. The position and tension between these three important considerations is constantly shifting.

Which scenario is healthier? The scenario on the left may represent a company that is performing relatively well with regard to contribution margin and the market but is suffering from a working capital crisis. The middle scenario depicts a company that is failing to generate cash and is suffering from a working capital crunch. The scenario on the right is a company that is generating a high amount of cash, has abundant working capital and a well-defended and growing base of customers.

But how can leadership best hope to manage these basic elements in a VUCA environment? The key to both the short-term and long-term management of these elements can be found in concepts called "coherence" and "resiliency."

Striving for Coherence and Resiliency

Coherence and resiliency are key terms in the emerging science of complex adaptive systems. What is a complex adaptive system (CAS)? First, let's understand that any complex system is governed by three important principles:

- Nonlinearity. Complex systems are best described as web connections, not linear connections. They loop and feedback on themselves interactively. The degree of complexity resulting from dynamic interactions can reach an enormous level. Dynamic interactions are explained as high degrees of inter-dependencies, non-linear interactions, short-range interactions, and positive and negative feedback loops of interactions.
- Extreme sensitivity to small initiating events. Lots of these initiating events occurring in a short time frame can produce significant nonlinear outcomes that may become extreme events. These events are often referred to as "lever point phenomena" or "butterfly events."
- Cause and effect are not proportional; a part that costs 10 cents can halt the assembly of multimillion dollar end items as quickly as a $10,000 part.[6]

The word "adaptive" introduces the element of how a complex system changes or reconfigures itself through a process known as "emergence." Once emergence has occurred, then feedback and selection occur over a period of time resulting in further reconfiguration to the system. When complex systems are co-mingled or intertwined (such as highly integrated supply chains) these events and steps tend to cascade across systems, making a highly complex and evolving picture. Figure 1-3 is a modified version of Figure 10.1 from the book *Demand Driven Performance—Using Smart Metrics*. It lists nine characteristics of a CAS.

CAS Key Characteristic	Explanation
Boundaries	All system boundaries are defined by their subsystems, hierarchies, adaptive agents, and signal sets to trigger action and interactions between them. No true boundary exists in a CAS because they are always part of a larger ecosystem but boundaries are practically defined by the limits at which the system's adaptive agents can act and affect change within.
Coherence	A CAS depends on its subsystems to align their purpose and actions with that of the system. Subsystems that are not in alignment can cause coherence to break down and push the system into chaos.
The Edge of Chaos	All CAS operate in a zone between stable equilibrium points and total randomness otherwise known as chaos. These zones have been learned or defined and are monitored by adaptive agents through sets of signals. Adaptive agents will act and even self-limit in order to attempt to keep the system out of chaos.
Self-Organization, Innovation and Emergence	CAS have emergent properties or events in which adaptive agents self-organize based on signal strength in order to solve an issue. Through feedback and selection they will learn and innovate in order to bring the system to an evolved state.
Signals	All CAS use defined signals to communicate with and within subsystems as well as with other systems that they interact with. System coherence and signal alignment is the primary consideration in determining the actual signals, their triggers and subsequent instruction sets.
Adaptive Agents	All CAS are dependent on adaptive agents to receive, interpret, and react to signals. They are adaptive because they are responsible for identifying and learning about changes in the patterns of signals and making recommendations for changes and innovation.
Signal Strength	All CAS depend on adaptive agents to recognize and prioritize across signals with system coherence always in mind.
Feedback Loops	All CAS depend on feedback loops to provide adaptive agents the ability to measure, sense, and adapt subsystem and system performance in order to maintain coherence and drive innovation and emergence.
Resilience	Resilience is the ability of a CAS to return to equilibrium after a large disturbance (either imposed or self-imposed). If a CAS is too rigid, the disturbance will push it into chaos.

FIGURE 1-3 Complex adaptive system characteristics[7]

At this time we will focus on only two of these characteristics. The first is coherence. A complex adaptive system's "success" depends on the coherence of all of its parts. A subsystem's purpose has to be in alignment with the purpose of the greater system in order for there to be coherence. Without that alignment, the subsystem acts in a way that endangers the greater system it depends on. Coherence must be at the forefront of determining the signal set components, triggers, and action priorities. To maintain coherence, adaptive agents must ensure their signal sets contain the relevant information to direct their actions and are not at cross-purposes with the goals of the systems it depends on.[8] The concept of coherence is consistent with the systemic approaches of thought leaders such as Ohno, Goldratt, and Deming and their respective disciplines of lean, Theory of Constraints, and Six Sigma. All of these disciplines urge management to organize and operate in a manner that emphasizes carefully aligning local actions to the global objective. Deming and Goldratt in particular were extremely opinionated about the failure of management and executives to understand and effectively embrace this concept, which is one that seems rooted in basic common sense.

The second characteristic of a CAS to be explored is resiliency. Resiliency allows a system to respond to a disturbance while maintaining equilibrium within its system boundary. In supply chain words, resilience is how well a system can return to stability when it experiences random or self-imposed variation. Resilience arises from the subsystem's ability to respond to the feedback loops that regulate equilibrium. The ability to adapt and the diversity or flexibility of options/actions determines how quickly the system can recover and/or improve. The opposite of resiliency is rigidity.[9]

Obviously, if a system is insufficiently resilient relative to the level of disturbance, it is at risk of collapse. Reeves, Levin, and Ueda identified six basic risks to the resiliency of a complex system.[10] Any organization wishing to avoid the threat of collapse must mitigate these risks. In the VUCA world these risks are more prevalent than ever.

- **The COLLAPSE risk.** A change from within or outside the industry renders the firm's business model obsolete. An obvious example is the impact that the emergence of online retailing giants such as Amazon had on the retail industry. At one time Sears and Roebuck was the largest retailer on the planet, headquartered in the tallest building in the western hemisphere, The Sears Tower. Sears initially built its empire through catalog sales, selling hardware, appliances, tools, and even plans for homes (The Craftsman). By 1980 the vast majority of the United States population had a Sears store or outlet within an hour drive. At the time of this writing Sears is but a shell of what it had been, struggling to meet cash commitments and desperately trying to find a way to survive. After being bought by Kmart to form Sears Holding, the Sears Tower is now the Willis Tower. The building is no longer the tallest in the western hemisphere and United Airlines now occupies much of the building. Sears Holding continues to sell brands and close stores from its location in Hoffman Estates, IL, a suburb of Chicago.

- **The CONTAGION risk.** Shocks in one part of the business spread rapidly to other parts of the business. In 2018 Ford had to close its profitable F150 assembly plant due to a fire in a Chinese-owned supplier located in Michigan. The fire affected many auto suppliers, but the hardest hit was Ford, specific to the F150 truck. The F150 is a multibillion dollar brand for Ford and substantially drives Ford's profits.
- **The FAT-TAIL risk.** Rare but large shocks, such as natural disasters, terrorism, and political turmoil. Examples here include the tsunami in Japan that affected the automotive and electronics industries. September 11, 2001 stopped industry across the United States and many companies never recovered, especially the airline industry. Massive consolidation and cutbacks in flights and service redefines the new airline industry.
- **The DISCONTINUITY risk.** The business environment evolves abruptly in ways that are difficult to predict. The financial crisis of 2008 created the biggest disruption to the U.S. housing market since the Great Depression. Increased foreclosure rates in

2006 and 2007 led to a banking crisis in 2008. Concerns about the impact of the collapsing housing and credit markets on the general U.S. economy caused the U.S. President to announce a limited bailout of the country's housing market for homeowners who were unable to pay their mortgage debt. This spilled over into other markets. People simply did not have money to spend. The automotive industry was shocked by the bankruptcy of General Motors. GM was the world's largest car maker and now it faced collapse because it no longer had sufficient cash to continue operation. It took the U.S. government stepping in to save a national icon and the jobs associated with it.

- **The OBSOLESCENCE risk.** The enterprise fails to adapt to changing consumer needs, competitive innovations, or altered circumstances. Blackberry was the first "smart" phone on the market. Market acceptance of this innovative device that could do email, phone calls, Internet, and a variety of other tasks, however, was leapfrogged by Apple's iPhone innovation. Blackberry quickly became viewed as obsolete as it lacked a new visual intuitive user interface as well as the access to thousands of specific "apps."

- **The REJECTION risk.** Participants in the business's ecosystem reject the business as a partner. The impact of social media has dramatically increased this type of risk. In 2017 a passenger filmed United Airlines personnel forcibly dragging a bloodied passenger off one of its planes. The video went viral, prompting an outcry in both the United States and China (the passenger was of Chinese descent). The airline posted a profit plunge of almost 70 percent in that quarter.

With the exploration of these two characteristics (coherence and resiliency) of a CAS, consider two pivotal questions:

- With regard to coherence: what is the goal of a for-profit company and how can the subsystems' purposes be best aligned to that goal? The three vital metrics of contribution margin, working capital, and customer base are far too remote from the subsystem's decisions to make them the metric's focus at the subsystem level. Is there some concept or principle that can ensure those three basic necessities at the higher level but that translates effectively all the way down to and through the subsystems and their respective operational levels? Without this answer, maintaining coherence is under constant threat. This question will be answered in the next section of this chapter.

• With regard to resilience: where is the starting point for an organization to create a framework to best mitigate these six risks? The exploration of an answer will begin in Chapter 2.

Authors' note: This is an extremely abbreviated description of complex adaptive systems. Readers seeking a deeper dive behind this science should consider reading Chapter 10 of *Demand Driven Performance —Using Smart Metrics* (Smith and Smith, McGraw-Hill, 2014) and the additional resources listed in that chapter.

Flow as an Objective and Purpose for Systems and Subsystems

What is the objective for every organization and a purpose for its subsystems to effectively tie to and drive that objective? Is there a basic fundamental principle to focus every business?

Now more than ever business is a bewildering and distracting variety of products, services, materials, technologies, machines, and people skills obscuring the underlying elegance and simplicity of it as a process. The required orchestration, coordination, and synchronization is simply a means to an end. That much is quite easy to grasp. What is more difficult for many organizations to grasp is what fundamental principle should underlie that orchestration, coordination, and synchronization.

The essence of any business is about flow. The flow of materials and/ or services from suppliers, perhaps through multiple manufacturing plants and then through delivery channels to customers. The flow of information to all parties about what is planned and required, what is happening, what has happened, and what should happen. The flow of cash back from the market to and through the suppliers.

Is this some sort of inspired revelation? No. Flow has *always* been the primary purpose of most services and supply chains. Very simply put, you must take things or concepts, convert or assemble them into different things or offerings and then get those new things or offerings to a point where others are willing to pay you for them. The faster you can make, move, and deliver all things and offerings, the better the performance of the organization tends to be. This incredibly simple concept is best described in what is known as Plossl's Law.

Plossl's Law

George Plossl was an instrumental figure in the formation and proliferation of Material Requirements Planning (MRP), the original planning and information system that would eventually evolve into modern-day Enterprise Resource Planning (ERP) products. He is commonly referred to as one of three founding fathers of these manufacturing systems; Joe Orlicky and Oliver Wight are the other two.

In 1975 Joe Orlicky wrote the book *Material Requirements Planning—The New Way of Life in Production and Inventory Management*. This book became the blueprint for commercial software products and practitioners alike. By 1981, led by Oliver Wight, Material Requirements Planning had evolved into Manufacturing Resources Planning (commonly referred to as MRP II). MRP II went well beyond the simple planning calculations used in MRP. It incorporated and used a Master Production Schedule, capacity planning and scheduling, and accounting (costing) data. Oliver Wight passed away in 1983 and Orlicky passed away in 1986. In 1994, George Plossl took up the torch to paint a vision for the future in the second edition of Joe's seminal work, *Orlicky's Material Requirements Planning* 2e.

Nineteen ninety-four, however, was an interesting and difficult time to write a book about the vision and implementation of technology. Mainframes had given way to client-server configurations and the birth of the Internet had people puzzled about its industrial application. But Plossl successfully navigated these difficulties by sticking to a key transcendent principle, one that had not been well articulated previously in manufacturing systems literature. He called it the First Law of Manufacturing. We now simply know it as Plossl's Law.

All benefits will be directly related to the speed of flow of information and materials.

It should be noted that Plossl was focused on manufacturing centric entities. As such there is an obvious omission to the law about services. But providing services is also all about flow. For example, the process of obtaining a mortgage flows through a series of steps. The longer it takes, the more risk to the sale and more dissatisfied the customer. Having a surgical procedure at a hospital is about flow. The longer the process, the higher the cost and risk to the patient. Repairing a downed piece of equipment is about flow. The longer the repair takes, the less revenue that piece of equipment can generate. As such, consider an amended Plossl's Law.

All benefits will be directly related to the speed of flow of information, materials, and services.

This statement is not just simple; it is elegant. It also requires a critical caveat in the modern services and supply chain landscape. We will get to that critical caveat in due time. First, let's further explore the substance of Plossl's Law.

"All benefits" is quite an encompassing statement. It can be broken down into components that most companies measure and emphasize. All benefits encompass:

- **Service.** A system that has good information and material and/ or services flow produces consistent and reliable results. Most markets and customers have an appreciation for consistency and reliability. Consistency and reliability are key for meeting customer expectations, not only for delivery performance, but also for things like quality. This is especially true for industries that have shelf-life issues and erratic or volatile demand patterns.

- **Quality.** When things are flowing well, fewer mistakes are made due to less confusion and expediting. This is not to say that qualities issues will not occasionally happen, but what it does say is that quality issues related to poor flow will most likely be minimized. This is important in industries with large assemblies with deep and complex bills of material and complicated routings to be scheduled. Frequent and chronic shortages cause work to be set aside to wait for parts, creating large work-in-process queues and then the inevitable expediting to get the work through the system.

- **Revenue.** When service and quality are consistently high, a company is afforded the opportunity to better exploit the total market potential. This means higher revenue volume from both the protection and growth of margin and market share.

- **Inventories.** With good flow purchased, work-in-process (WIP) and end item inventories will be minimized and directly proportional to the amount of time it takes to flow between stages and through the total system. The less time it takes products to move through the system, the less the total inventory investment. The simple equation is Throughput $\times$ Lead Time $=$ WIP. Throughput is the rate at which material is exiting the system. Lead time is the time it takes to move through the system and WIP is the amount of inventory contained between entry and exit. A key assumption is that the material entering the system is proportionate to the amount exiting the system. The basis for this equation is the queuing theory known as Little's Law.

It is also worth noting that to maintain flow, inventories cannot be eliminated. Flow requires at least a minimal amount of inventory. Too little inventory disrupts flow and too much inventory also disrupts flow. Thus, when a system is flowing well, inventories will be "right-sized" for that level of flow. What we will find out later is that the placement and composition of inventory queues will be a critical determinant in how well flow is protected and promoted and what "right-sized" really means in terms of quantity and working capital commitment.

- **Expenses.** When flow is poor, additional activities and expenses are incurred to correct or augment flow problems. In the short term it could mean expedited freight, overtime, rework, cross-shipping, and unplanned partial ships. In the longer term it could mean additional and redundant resources and third-party capacity and/ or storage. These additional short- and long-term efforts and activities to supplement flow are indicative of an inefficient overall system and directly leads to cash exiting the organization.

- **Cash.** When flow is maximized, the material that a company paid for is converted to cash at a relatively quick and consistent rate. Additionally, the expedite-related expenses previously mentioned are minimized, reducing cash unnecessarily leaving the organization. This makes cash flow much easier to manage and predict and will also lead to less borrowing related expenses.

Furthermore, the concept of flow is also crucial for project management. R&D and innovation efforts that flow well can impact and amplify all of the above benefits as the company exploits these efforts.

What critical business equation is defined with these six basic benefits? It is an equation that defines and measures the very purpose of *every* for-profit organization: ***to protect and grow shareholder equity.*** This is and always has been the basic responsibility and duty of every executive.

In its simplest form the equation to quantify this purpose is:

$$\text{Net Profit} \div \text{Investment} = \text{Return on Investment}$$

Net profit is a company's revenue, which consists of total sales dollars collected through a particular period subtracting the operating expenses, cost of goods sold (COGS), and interest and taxes within that same period. Investment is simply the captured money in the system needed to produce the output. The simplicity of this formula can make it easy to manipulate depending on how one defines time periods, but essentially it is a measure of the money that can be returned by a system versus the money it takes to start and maintain that system. Thus, the output of the equation is called return on investment. The higher the rate of return on investment (both in the short run and the anticipated long run), the more valuable the shareholder equity.

Of course, a full DuPont analysis would provide a more detailed perspective incorporating profit margin, asset turnover, and financial leverage. The above equation is simply a conceptual shortcut that can be used to make a crucial connection between flow and return on investment via Plossl's Law.

Hopefully, that connection is now readily apparent. The previously mentioned benefits of flow (perhaps with the exclusion of taxes) are all direct inputs into the ROI equation. This makes flow *the* single biggest lever in determining *the* objective of a for-profit organization. This can be expressed as the equation in Figure 1-4. This depiction first appeared in the book *Demand Driven Performance—Using Smart Metrics* (Smith and Smith, McGraw-Hill, 2014, p. 71).

$$\Delta\text{Flow} \rightarrow \Delta\text{Cash Velocity} \rightarrow \Delta \left(\frac{\text{Net Profit}}{\text{Investment}} \right) \rightarrow \Delta\text{ROI}$$

FIGURE 1-4   Connecting flow to return on investment (ROI)

Explaining this equation requires first a definition of the elements and then how they relate to each other.

- **Flow.** The rate at which a system converts material to product **required by a customer**. If the customer does not want or take the product, then that output does not count as flow. It is retained in the system as investment (captured money).

- **Cash velocity.** The rate of **net cash generation**; sales dollars minus truly variable costs (aka contribution margin) minus period operating expenses.
- **Net profit/investment.** Net profit divided by investment (captured money) is the equation for ROI.

The delta and yield arrows in the equation explain the relationships between the components of the equation. Changes to flow directly yield changes to cash velocity in the same direction. As flow increases so does cash velocity. Conversely, as flow decreases so does cash velocity. As cash velocity increases so does return on investment as the system is converting materials to cash more quickly.

When cash velocity slows down, the conversion of materials to cash slows down. The organization is simply accomplishing less with more. This scenario typically results in additional cash velocity issues related to expediting expenses. Period expenses rise (overtime) or variable costs increase (fast freight, additional freight, and expedite fees). This directly reduces the net profit potential within the period and thus further erodes return on investment performance.

The River Analogy

The simple analogy to this equation is the manner in which a river works. Water flows in a river as an autonomous response to gravity. The steeper the slope of the river bed, the faster the water flows. Additionally, the fewer number of obstructions in the river, the faster the water runs.

In service and supply chain management, materials and/or services flow through the network like water through a river. Materials are combined, converted, and then moved to points of consumption. Services are offered, scheduled, and delivered to customers. The autonomous response of these flows is demand. What else could it or should it be? Ideally, the stronger the demand, the faster the rate of flow of materials and services. And like rivers, service and supply chains have obstructions or blockages created by variability, volatility, and limitations in the "river bed." Machines break down, critical components are often unavailable, yield problems occur, choke points exist, capacity bottlenecks exist, etc. All of these issues are simply impediments to flow and result in "pools" of inventory with varying depth. A river without flow is not a river, it is a lake. Operations with out flow is a disaster.

With this analogy we begin to realize that flow is the very essence of why the Operations subsystem of manufacturing and supply chain companies even exist. The Operations subsystem is typically divided into functions, each of which have a primary objective for which they are responsible and accountable. Figure 1-5 is a simple table showing typical Operations functions and their primary objective.

Operational Function	Primary Objective
Planning	Synchronize supply and demand
Logistics	Connect sources to consumption points
Purchasing	Ensure material/component availability
Shop-Floor	Execute the schedule
Scheduling	Sequence activity to meet commitments
Quality	Meet specification

FIGURE 1-5   Typical functions in operations

All of these objectives are protected and promoted by encouraging flow. Under what scenario does a cost-based focus enable you to synchronize supply and demand or sequence activity to meet commitments? In fact, a cost-based focus most often leads to the exact opposite of these objectives. Thus, if Operations and its functions want to succeed in being truly effective, there is really only one focus—*flow*. Flow must become the common framework for communications, metrics, and decision making in Operations.

Let's expand this view to the organization as a whole. An organization is typically divided into many subsystems, not just Operations. Each subsystem is typically tasked with its own primary objective. Figure 1-6 is another simple table showing the typical subsystems of a manufacturing and/or supply chain-centric company.

Organizational Function	Primary Objective
Finance	Drive shareholder equity
Sales	Capture demand
Marketing	Create brand awareness and demand
Operations	Utilize assets to meet demand
Engineering/R&D	Drive innovation

FIGURE 1-6   Typical organizational functions

All of these functional objectives require flow to be promoted and protected to drive maximum effectiveness. When things are flowing well, shareholder equity, sales performance, market awareness, asset utilization, and innovation are promoted and protected and costs are under control. This was discussed extensively earlier in this chapter. Thus, flow must become the common framework for communications, metrics, and decision making across the organization.

Additionally, flow is also a unifying theme within most major process improvement disciplines and their respective primary objectives.

- Theory of Constraints (Goldratt) and its objective of driving system throughput
- Lean (Ohno) and its objective to reduce waste
- Six Sigma (Deming) and its objective to reduce variability

All of these objectives are advanced by focusing on flow first and foremost. This should not be surprising since we have already mentioned these thought leaders regarding systemic coherence. When considered with Plossl's First Law of Manufacturing, the convergence of ideas around flow is quite staggering. There should be little patience for ideological battles and turf wars between these improvement disciplines; it is a complete waste of time. All need the same thing to achieve their desired goal: flow. Among these disciplines, flow becomes the common objective from a common strategy based on simple common sense grounded in basic physics, economic principles, and complex systems science.

The concept and power of flow is not new, but today it seems almost an inconvenient afterthought that managers must, if pressed, acknowledge as important. It powered the rise of industrial giants and gave us much of the corporate structure in use today. Leaders such as Henry Ford, F. Donaldson Brown, and Frederick Taylor made it the basis for strategy and management. The authors believe these leaders would be astonished at how off the mark modern companies are when it comes to flow; they are surviving in spite of themselves.

Thus, Plossl's Law, while incredibly simple, should not be taken lightly. This one little statement has *always* defined the way to drive shareholder equity and it was articulated by one of the main architects of conventional planning systems! Embracing flow is the key, not to just surviving but adapting, taking a leadership position or being a fierce and dangerous competitor. It is the first step to becoming a Demand Driven Adaptive Enterprise. In order for that first step to be taken, however, a huge obstacle must be overcome: the universal fixation, emphasis, and obsession over cost.

Cost and Flow

Executives of corporations around the world obsess over cost performance, most particularly unit cost. It dominates discussions on a daily basis and constantly influences a majority of strategic, tactical, and operational decisions throughout the organization. We need to understand what unit cost is (and is not) and why it became so prominent.

Any unitized cost calculation has always been based on past activity within a certain period. The calculation of standard unit cost attempts to assign a cost to an individual product and/or resource based on volume and rate over a particular time period. Essentially, fixed and variable expenses within a period are accumulated and divided by volume within that period to produce a cost per unit. This cost per unit can also be calculated by resource and location. This cost then becomes a foundation for many metrics and decisions at the operational, tactical, and strategic levels in the present and for the future.

Where did unit cost come from? In short, unit cost can trace its origin from the 1934 adoption of Generally Accepted Accounting Principles (GAAP) by the United States government as an answer to the U.S. stock market crash of 1929. Many industrialized countries have since followed this example. GAAP is an imposed requirement for the fair and consistent presentation of financial statements to external users (typically shareholders, regulatory agencies, and tax collection entities). GAAP reporting and the unitized cost calculations based on it was then incorporated into information systems circa 1980 with the advent of manufacturing resources planning (MRP II), the pre-cursor to enterprise resource planning (ERP). That incorporation continues today in every major ERP system.

Why was GAAP incorporated into information systems? It was not driven by the need to manage cost today, or to make management decisions or develop strategy in the future; it was driven by the need to fulfill the financial reporting requirements of GAAP in a much easier and quicker fashion. Even today most ERP implementations begin with the financial module. In the United States the Sarbanes-Oxley act of 2002 drove ERP software companies to provide technology that allowed even faster financial reporting using these rules.

GAAP, however, does not and should not care about providing internal management reporting. Why? Because GAAP's entire purpose is to provide a consistent reporting picture about what happened over a past period, *not* what should be done today or suggestions or predictions for the future. GAAP is simply a forensic snapshot of past performance within a certain defined time period, meaning that if it is done as required by the law, it is always 100 percent accurate in determining past cost performance information only.

The incorporation of these cost data quickly led to its numbers and equations becoming the default way to judge performance and make future decisions. Why? The higher-level metrics like return on investment, contribution margin, working capital, etc. are much too remote to drive through the organization. Management needed something to drive down through the organization to measure performance. Cost numbers were readily available and constantly updated in the system.

Now, unfortunately, most of management actually believes or accepts that these numbers are a true representation of cash or potential. The assigned standard fixed cost rate, coupled with the failure to understand the basic aspects of a complex system leads managers to believe that every resource minute saved anywhere is computed as a dollar cost savings to the company. GAAP unit costs are used to estimate both cost improvement opportunities and cost savings for batching decisions, improvement initiatives and capital acquisition justifications. In reality the "cost" being saved has no relationship to cash expended or generated and will not result in an ROI gain of the magnitude expected. Cost savings are being grossly overstated.[11]

In 2018 a joint study released by APICS and the Institute of Management Accountants named three significant issues regarding costing information that supply chain professionals receive from costing systems.

- "An overreliance on external financial reporting systems: many organizations rely on externally-oriented financial accounting systems that employ oversimplified methods of costing products and services to produce information supporting internal business decision making.
- Using outdated costing models: traditional cost accounting practices can no longer meet the challenges of today's business environment but are still used by many accountants.
- Accounting and finance's resistance to change: With little pressure from managers who use accounting information to improve data accuracy and relevance, accountants are reluctant to promote new, more appropriate practices within their organizations."[12]

It is the job of management accounting, which is a different profession with an entirely different body of knowledge than financial cost accounting, to provide meaningful and relevant information for decision making. While the body of knowledge still exists there are few with real deep expertise in it. What happened to all the management accountants? They were essentially stripped out of mid-management in the 1980s because they were deemed largely irrelevant because executives believed that the system could now effectively tell the organization what to do via the automated cost data.

Two important points must be made about cost at this point. First, any measurement based on past activity is guaranteed to be wrong in the future. Assuming that past cost performance will be indicative of future cost performance in the VUCA environment is simply nonsensical.

Second and most importantly, good flow control actually yields the best cost control. If things flow well within a period, the previously described benefits of flow occur during that period. If those previously described benefits happen, then fixed (depending on the length of the period) and variable expenses are effectively controlled in combination with better volume performance. This will be reflected in the GAAP statements produced for that entire past period. Thus, emphasizing flow will actually be more effective even for cost accountants!

This leads us to an interesting yet simple rule about cost and flow, a corollary to Plossl's Law:

> When a business focuses on flow performance, better cost performance will follow. The opposite, however, is not the case.

It should be noted that embracing a flow-based focus is not license or a strategy to overspend on massive amounts of capacity, constantly employ overtime, and expedite everything. That is not a flow-based focus. Those tactics become necessary mainly because a company is not primarily focused on flow.

In Chapter 2 we will dive into this corollary in more depth. At this point, however, the conclusion should be relatively solid: if a business wants to manage cost performance it must first and foremost design and manage to flow performance. The mistake to use GAAP-generated cost numbers and metrics as operational tools is actually a self-imposed limitation by an organization's management. But what can we use in its place? It has to be something that emphasizes flow now and in the future.

With this flow and systems perspective there are at least two additional corollaries to Plossl's Law that are worth mentioning at this point:

> Something is productive if and only if it leads to better promotion and protection of system flow.

Something is deemed efficient if and only if it leads to better promotion and protection of system flow.

Variability and Flow

To fundamentally understand how to emphasize flow now and in the future, we must first understand the biggest determinant in the management of flow: the effective management of variability. In Figure 1-7 we see an expanded form of the equation previously introduced. Variability is defined as the summation of the differences between our plan and what actually happens. As variability rises in an environment, flow is directly impeded. Conversely, as variability decreases, flow improves.

$$\Delta \text{Variability} \rightarrow \Delta \text{Flow} \rightarrow \Delta \text{Cash Velocity} \rightarrow \Delta \left(\frac{\text{Net Profit}}{\text{Investment}} \right) \rightarrow \Delta \text{ROI}$$

FIGURE 1-7 Connecting variability to the flow equation

The impact of variability must be better understood at the systemic rather than the discrete detailed process level. The war on variability that has waged for decades has most often been focused at a discrete process level with little focus or identified impact to the total system; Deming would not be pleased. Variability at a local level in and of itself does not necessarily impede system flow. What impedes system flow is the accumulation and amplification of variability across a system. Accumulation and amplification happens due to the nature of the system, the manner in which the discrete areas and environment interact (or fail to interact) with each other. Remember the three characteristics of complex systems: nonlinearity, extreme sensitivity to small initiating events, and a disproportion between cause and effect. Smith and Smith proposed the Law of System Variability.

The more that variability is passed between discrete areas, steps or processes in a system, the less productive that system will be; the more areas, steps or processes and connections between them, the more erosive the effect to system productivity.[13]

Quite simply, Figure 1-7 says that when things don't go according to plan, flow is directly impacted. Is this really surprising? Methods like Six Sigma, lean and Theory of Constraints have recognized the need to control variability for decades. Unfortunately, many of those methods point to or get focused on limited components or subsystems of an organization or supply chain. Most of them attempt to compensate for variability *after* a plan has been developed and implemented (a plan that is typically built utilizing a design that assumes everything will go according to plan—an extremely poor assumption).

The Rise of VUCA

The world is a much different place today than it was 50 years ago, when the conventional operational rules and systems were developed. Figure 1-8 is a list of some dramatic changes in supply chain–related circumstances that have occurred since 1965.

Supply Chain Characteristics	1965	Today
Supply Chain Complexity	Low	High
Product Life Cycles	Long	Short
Customer Tolerance Times	Long	Short
Product Complexity	Low	High
Product Customization	Low	High
Product Variety	Low	High
Long Lead Time Parts	Few	Many
Forecast Accuracy	High	Low
Pressure for Leaner Inventories	Low	High
Transactional Friction	High	Low

FIGURE 1-8   Supply chain circumstances in 1965 versus today

The circumstances under which Orlicky and his cadre developed the rules behind MRP and surrounding techniques have dramatically changed. Customer tolerance times have shrunk dramatically, driven by low informational and transactional friction largely due to the Internet. Customers can now easily find what they want at a price they are willing to pay for it and get it in a short period of time.

Ironically, much of this complexity is largely self-induced in the face of these shorter customer tolerance times. Most companies have made strategic decisions that have directly made it much harder for them to effectively do business. Product variety has risen dramatically. Supply chains have extended around the world driven by low cost sourcing. Product complexity has risen. Outsourcing is more prevalent. Product life and development cycles have reduced.

This has served to create a huge disconnect between customer expectations and the reality of what it takes to fulfill those expectations reliably and consistently. This will not get better any time soon. The proliferation of quicker delivery methods such as drones will simply serve to widen the disparity between customer tolerance time and the procurement, manufacturing, and distribution cycle times. Many supply chains are ill prepared for this storm intensifying.

Add to this an increased amount of regulatory requirements for consumer safety and environmental protection and there are simply more complex planning and supply scenarios than ever before. The complexity comes from multiple directions: ownership, the market, engineering, and sales and the supply base. Ultimately, this complexity manifests itself with a high degree of volatility, uncertainty, and ambiguity. It is making it much more difficult to generate realistic plans and maintain the expectation that things will go according to plan, especially when those plans are based on GAAP-derived drivers.

The key to protecting and promoting flow at the system level is to understand and manage variability at the system level. What then is the key to managing variability? In order to answer that question we will need to expose another key component of the flow equation, the component that eludes most companies in today's complex and volatile supply chain environments.

Relevance Found

There is an important factor in managing variability that must be recognized; without it, the quest to reduce or manage variability at the systemic level is a quixotic one at best. This missing element is labeled as "Visibility" in Figure 1-9.

$$\Delta \text{Visibility} \rightarrow \Delta \text{Variability} \rightarrow \Delta \text{Flow} \rightarrow \Delta \text{Cash Velocity} \rightarrow \Delta \left(\frac{\text{Net Profit}}{\text{Investment}} \right) \rightarrow \Delta \text{ROI}$$

FIGURE 1-9 Adding visibility to the equation

Visibility is defined simply as access to relevant information for decision making.[14] This provides an extremely important caveat to Plossl's Law. A company cannot just indiscriminately move data and materials quickly through a system and expect to be successful. Today organizations are frequently drowning in oceans of data with little relevant information and large stocks of irrelevant materials (too much of the wrong stuff) and not enough relevant materials (too little of the right stuff). Furthermore, sophisticated real time analytics of bigger and bigger databases will not solve the problem but instead will create a deeper, wider, and stormier ocean of data and materials unless we understand how to better sift through that ocean to determine what is truly relevant for decision making now and in the future.

Note that this formula now starts not with flow but with what makes information relevant. If we don't fundamentally grasp how to generate and use relevant information, then we cannot hope to manage variability and consequently facilitate flow. Moreover, if we are actively blocked from generating or using relevant information by systems, then even if people (adaptive agents) understood there was a problem, they would be essentially powerless to do much about it.

What makes the flow of information, materials, and services relevant is its relationship to the required output or market expectation of the system now and in the future, *not* what was accomplished (or not accomplished) in the past. To be relevant, the information, materials, and services must synchronize the assets of a business to what the market really wants now and in the future; no more, no less.

Thus, we have reached the core problem plaguing most organizations today: the inability to generate and use relevant information to effectively manage variability to then protect and promote flow and drive ROI performance. Without addressing this core problem, there can be no systemic solution for flow.[15] Figure 1-10 shows the core problem area of the equation versus the area associated with Plossl's Law as first stated.

FIGURE 1-10   Core problem area of the equation[16]

Having visibility to the right information is a prerequisite to effectively managing variability and ensuring the flow of the right materials at the right time. With this is mind, Plossl's Law now must be amended to:

*All benefits will be directly related to the speed of flow of **relevant** information, materials, and services.*

A New Appreciation for the Bullwhip Effect

But this core problem is not confined to individual organizations. As discussed previously, complex systems (organizations) interact collectively with other complex systems to create an even larger complex system. What is happening at this higher level?

There is a phenomenon involving the stated core problem that dominates most supply chains and complex systems. This phenomenon is called the "bullwhip effect." The fourteenth edition of the APICS dictionary defines the bullwhip effect as:

"An extreme change in the supply position upstream in a supply chain generated by a small change in demand downstream in the supply chain. Inventory can quickly move from being backordered to being excess. This is caused by the serial nature of communicating orders up the chain with the inherent transportation delays of moving product down the chain. The bullwhip can be eliminated by synchronizing the supply chain."[17]

A massive amount of research and literature has been devoted to the phenomenon known as the bullwhip effect, starting with Jay Forrester in 1961. However, very little, if any, of that body of knowledge has been devoted specifically to its bidirectional nature. Most of the research has been dedicated to understanding how and why demand signal distortion occurs and how to potentially fix it through better forecasting algorithms, tightly synchronizing the supply chain.

The bullwhip is really the systematic and bidirectional breakdown of information and materials in a supply chain. Figure 1-11 is a graphical depiction of the bullwhip effect. Distortions to relevant information go up the chain, growing in size and causing wider and wider oscillations both in terms of quantity and timing requirements. The wavy arrow moving from right to left represents that distortion to relevant information in the supply chain. The arrow wave grows in amplitude in order to depict that the farther up the chain you go, the more disconnected the information becomes from the origin of the signal as signal distortion is transferred and amplified at each connection point.

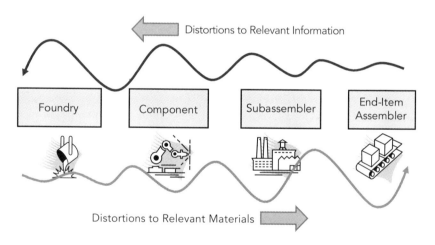

FIGURE 1-11 Illustrating the bidirectional bullwhip effect

Distortions to relevant materials come down the chain as delays and shortages accumulate. The more connections that exist, the more pronounced the delays and shortages. The wavy arrow from left to right depicts that distortion. Lead times expand, shortages are more frequent, expedites are common, and flow breaks down.

Summary

Creating visibility to relevant information and managing the risks to coherence and resiliency is no trivial task, but it is the only path to sustainable organizational success as measured by ROI. The more relevant information our organization has, the more immediate and enduring success it will have—it is really that elementary. Intuitively, people in organizations know that they must find and use relevant information for decision making. Yet many of those people recognize that their information systems are not giving them the visibility that they need. What will it take to get more relevant information throughout the organization? This question is explored in Chapter 2.

The Prerequisites for Relevant Information

The search for and use of relevant information to control variability, promote flow, and ultimately drive ROI begins with an understanding of four basic prerequisites. The absence of one or more of these prerequisites will dramatically compromise the integrity of the relevance of the information.

Prerequisite #1: Understanding Relevant Ranges

The concept of relevant range comes from economics and management accounting. Relevant range refers to the time period in which assumptions are valid. Trying to force fit assumptions (and metrics derived from those assumptions) into an inappropriate time range directly results in distortions to relevant information, creates variability, and thus causes a breakdown in the flow of relevant materials and services.

The assumptions and information that are valid and relevant within different time ranges will differ dramatically and these differing ranges are utilized by different personnel. For example:

- Forecasts are highly relevant in thinking about the future, they are essentially useless with regard to what needs to be done today.
- Fixed expenses are variable in the longer range, not in the close-in range (that is why they are called "fixed").

- Work order delays are relevant for the current schedule, not for an executive trying to determine the company's strategy for the next year.
- A machine breakdown is relevant for a maintenance crew's activity, not what the company thinks it will sell in Q3.

There are at least three distinct relevant ranges that we must deal with: operational, tactical, and strategic. The specific time value of each relevant range will differ between companies depending on the circumstances involved in that company. Chapter 3 will describe a way to discretely determine these ranges for every organization.

Prerequisite #2: Tactical Reconciliation Between Relevant Ranges

While the assumptions and information that are relevant for decision making differ between ranges, there is still an absolute need to reconcile these differences on an ongoing and iterative basis; the signals at all levels must align and reconcile. Strategy must be influenced by defined operational capability and performance as well, considering how the operating model might perform under additional predicted scenarios. Operational capability must be influenced by predicted scenarios and/ or strategic expectations in future time periods to ensure that the operational capability supports the strategy. But how does this reconciliation actually occur in a true bidirectional fashion?

Between the strategic and operational ranges is the tactical range. As we will see in Chapters 3 and 5, this tactical range is unique, and for it to perform the necessary reconciliation it must become a bidirectional reconciliation hub for the key characteristics of a CAS.

Prerequisite #3: A Flow-Based Operating Model

A flow-based operating model is an operating model specifically designed on the fundamental principle of flow. Chapter 1 firmly established two things:

- Promoting and protecting flow is the key to increased return on investment. This is the very essence of Plossl's Law.
- Flow can be a unifying factor, not just in operations but also between functions, to advance their individual primary objectives.

A flow-based operating model makes the following critical assumptions:

- Assets must be as closely synchronized to actual demand as possible. The cost of being wrong has grown dramatically with the rise of complexity and volatility. Synchronizing assets as close as possible to actual demand minimizes that risk.
- Variability cannot be eliminated but its impact on system flow can be effectively controlled and mitigated with the right operating model design. Every process, even when under statistical control, still experiences variability within its control limits; that is a given. The accumulation of this variability between processes is really the only relevant thing when it comes to managing system performance.
- Focusing on precisely synchronized planning and scheduling across all materials and resources is largely wasted effort. Most MRP and shop-floor schedules are unrealistic as soon as they are released. Constantly rescheduling to attempt to account for variation actually introduces additional variability and subsequent confusion.
- Some level of inventory is required at some stage in the supply chain because the customer tolerance time is insufficient to procure and make everything to order. Inventory cannot be entirely eliminated. At the same time, inventory should not be everywhere. The choice of where to place inventory is highly strategic since it determines the customer lead time and the level of working capital.
- Batching decisions should be based on flow considerations instead of cost considerations. Batches are an actual bona fide limita-

tion for most companies. How those batch sizes are determined, however, is a choice. Unfortunately, the vast majority of batching decisions are driven by cost performance metrics instead of flow-based determinants.

- Some level of protective capacity must exist in any environment with multiple interacting resources. Resources have disparate levels of capacity relative to their respective required tasks and volume. The perfectly balanced line is still a mythical creature with an enormous price tag. This disparity means that precaution must be taken against misusing or wasting points of additional capacity.

- Inventory is an asset and should be treated as any other asset on the balance sheet. Some lean enthusiasts like to promote that inventory is a liability; however, whether inventory is an asset or a liability depends on the position and quantity of that inventory. Production assets such as machines are added only when there is a positive return on investment. Inventory targets must be the result of a strategic choice for return on investment as well.

- Capability must be established based on the expected future for the company. Since it is not possible to predict the future precisely, the capability range is established by the combination of inventory placement and protective capacity. The size of this range is dependent on the uncertainty of the future. The establishment of this range is dependent on a comprehensive risk assessment of the market and supply chain.

- The objective of the operating model is to maximize margin by focusing on increasing service levels, enable premium pricing by providing a unique competitive market position, leverage constrained resources, and identify available capacity that can be used for incremental margin positive opportunities. This is in contrast to the more common expectation that the operating model should be developed to minimize unit cost and inventory while maximizing efficiency and utilization.

Thus, the question becomes: is there a well-defined and proven flow-based operational model that can scale to the multi-echelon enterprise level? Chapter 4 will focus on that model—the Demand Driven Operating Model.

Prerequisite #4: Flow-Based Metrics

If flow of relevant information, material, and services becomes the common focus for decision making in day to day operations and throughout the organization, then we need an appropriate suite of metrics for each of the relevant time ranges to support that focus. Appropriate metrics will allow us to maintain organizational coherence to that ROI objective now and in the future. Flow-based metrics should neither be a source of variability nor exacerbate variability.

Force fitting or failing to remove non-flow-based metrics in the deployed metrics suite will directly lead to conflicts and distortions throughout the organization—it will obscure what is relevant! Obscuring what is relevant directly leads to more variability, which in turn directly inhibits the flow of relevant information, materials, and services. Worse yet, when we do this, we are doing it to ourselves; it is self-imposed variation that directly hurts our ROI performance and causes great stress to the organization's personnel.

Any suite of flow-based metrics must consider three additional prerequisites:

- The metrics must fit the appropriate range.
- The metrics must be reconcilable between ranges.
- The metrics must fit the flow-based operating model.

These four prerequisites for relevant information define what it means to think, communicate, and behave systemically—the only way to protect and promote flow. If an organization and its personnel do not have this thoughtware installed, then the flow of relevant information and materials will always be impeded to varying degrees, leading directly to poorer ROI performance. Thus, before companies invest significant amounts of money, time, and energy into new hardware and software solutions, they must first consider investing in the proper thoughtware in order to gain visibility to what is truly relevant. Technology can only provide value if it addresses a limitation that the company is facing in terms of achieving its objective.

Convention's Failure with the Four Prerequisites

Now that the four prerequisites for relevant information are established, let's turn our attention to the failure of convention to establish and properly use these prerequisites. First, let's describe what the typical conventional approach looks like.

The conventional approach to managing a company involves strategic, tactical, and operational perspectives. Strategy is deployed to the organization by a Sales and Operations Planning (S&OP) process. The S&OP process balances supply availability and demand requirements resulting in a Master Production Schedule (MPS). The MPS is essentially intended to be a tactical dampener to prevent the forecast variability from driving MRP directly due to the imbalance of load against capacity. In the process of preparing the MPS, S&OP integrated reconciliation performs a rough-cut capacity check. The only bidirectional interaction between the MPS and S&OP is an "aggregation—disaggregation" process in which disaggregated demand forecasts are rolled up from the item level to the product family level to create the new forward-looking production plan. Then the production plan is rolled down into item level requirements by date, and this is used as the gross requirements line for MRP calculations.

In summary, the MPS is a statement of what can and will be built recognizing available capacity. The MPS sends this forward-looking plan for the planning horizon to Material Requirements Planning (MRP). MRP calculates the necessary supply order generation dates and quantities necessary to synchronize to that plan through a level by level explosion. Orders are then released when required by a date that has been calculated precisely from this multi-level explosion process. Figure 2-1 illustrates the conventional approach.

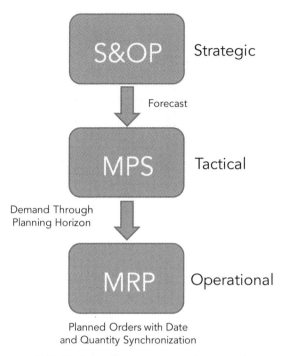

FIGURE 2-1   The conventional approach

Now we will turn our attention to the failure of convention regarding the four prerequisites to relevant information.

Convention and Relevant Ranges

The conventional approach clearly supports the need for relevant ranges. Figure 2-1 clearly shows a top-down linear approach that recognizes strategic, tactical, and operational emphases. The problem with convention, however, is that it improperly manages these ranges. This will be demonstrated through two examples that are devastating to relevant information and flow.

Our first example is conventional planning's reliance on forecasting item level demand for the planning horizon. Obviously, predicting market behavior and conditions is a necessary component for successfully managing any business. The better our forecasts, the better leadership can define and manage a company's path to success—there should be no doubt about that. However, bringing those predictions into the immediate operational range by tightly synchronizing order generation directly with a demand signal containing known error creates an immense amount of distortion and waste. There are three rules about forecasts:

- They start out wrong.
- The more remote in time the forecast is extended, the more wrong the forecast will be.
- The more detailed the forecast is, the more wrong the forecast will be.

Despite these well-known facts, convention continues to drive actual supply orders to forecast and then attempts to make corrections as better information is available. This means that capacity, capital, materials, and space are committed to signals that have significant rates of error associated with them. This is the very definition of irrelevant or at the very least distorted information and is one reason why forecasts are irrelevant in the short range. This is a clear mismanagement of the short term relevant range, demonstrating a lack of comprehension of this critical concept. Readers wishing to know more about this issue should consider reading the book *Precisely Wrong—Why Conventional Planning Fails and How to Fix It* (Ptak and Smith, Industrial Press, 2017).

Our second example of improperly managing relevant ranges is the use of fully absorbed unitized cost metrics for operational decisions. Fully absorbed unit cost means that all manufacturing costs, fixed and variable, are absorbed by the units produced. In other words, the cost of a finished unit in inventory will include direct materials, labor, and overhead costs.

Direct materials are variable costs. Variable cost is tied to unit volume, *not* resources. Variable costs rise and fall with unit volume but *do not* change on a per unit basis. Labor and overhead are fixed costs. Fixed costs are *not* affected by volume changes in activity level within the operational relevant range (a specific short-range period). Using fully absorbed unit cost related metrics directly creates the false impression that fixed costs can and will vary within the short range. They do not; that is why they are called fixed costs. The unitized cost equation obviously improperly mixes relevant ranges. This causes significant distortion in relevant information at the operational level and directly relates to disruptions in coherence and flow. Once again, this is a clear mismanagement of relevant range, demonstrating a lack of comprehension of this critical concept.

Convention and Tactical Reconciliation

In convention, tactical reconciliation is not bidirectional—it is essentially a one-way street in an assumed linear predictive world. This one-way direction limits the ability to drive any meaningful adaptation and additionally, any attempt at periodic reconciliation is incredibly painful. For example, every MRP run results in massive cascading schedule changes as date and quantity changes at higher levels affect all connected lower level components. This is an effect called nervousness and leaves most companies in a huge dilemma about how often they run MRP and how much they can actually trust its output, causing planners to export data to spreadsheets in an attempt to find the truth.

At a higher level, monthly S&OP updates create massive shifts at the beginning of every month that are then compounded by new MRP runs. Planning personnel struggle to stabilize the requirements picture and plans only to have the new version of the prescribed future reality dropped into their laps at the beginning of the next month. In convention, this makes tactical reconciliation more akin to a constant and repetitive cycle of tactical demolition and reconstruction; both are incredibly messy, wasteful, and cause great confusion.

Convention and Flow-Based Operating Models

There has certainly been no shortage of flow-based operating models proposed within the last fifty years.

The very purpose of Material Requirements Planning (MRP) was to perfectly synchronize supply and demand while netting inventory perfectly to zero. At a prima facie level that sounds great. In practice, like everything else that seems to be too good to be true, reality was far from expectations. As the gap between the total time it took to procure, produce, and deliver and the time the customer was willing to wait widened, the ability of MRP to perform as planned was further compromised. The MPS concept was developed in an attempt to synchronize the company around a single set of numbers and when reality exposed that master plan as a myth, MRP was still calculating requirements and generating supply orders with high rates of error.

Both the disciplines of lean and Theory of Constraints have proposed flow-based operating models for years. Lean proposes a flow-based model utilizing kanbans, supermarkets, and heijunka boards to react to actual orders in a pure pull approach. Theory of Constraints advocates yet another flow-based model using drum-buffer-rope or critical chain scheduling with time, capacity, and stock buffering. The authors have extensive experience in implementing both disciplines. Unfortunately, those models were incomplete in that they failed to sufficiently deal with information system challenges and adequately define alternatives at the enterprise level. As a result, the implementations of these flow-based models were typically compartmentalized within small subsystems with a limited life due to constant conflict interacting with other non-flow-based areas. Alternatively, these concepts were limited to smaller enterprises that used ad hoc tools to run the flow-based system.

Information systems are not and have never been the enemy of relevant information. The inappropriate rules underlying the implementation of those information systems are the real enemy. Simply put, those rules can be changed and/or eliminated for appropriate flow-based alternatives. The successful integration of information systems into a flow-based model is a must for sustainable success.

Convention and Flow-Based Metrics

Make no mistake, there are important flow-based metrics in use in conventional approaches. Metrics like on-time delivery, inventory turns, and fill rates are flow-based. Their use, however, is countered by conflicting cost-based metrics such as fully absorbed unit cost and operational equipment effectiveness (OEE) at all operations. These conflicting metrics obscure what is relevant and introduce significant self-imposed variability within organizations as personnel oscillate between protecting flow and protecting cost performance. This is a huge threat to coherence and incredibly stressful to the organization's personnel.

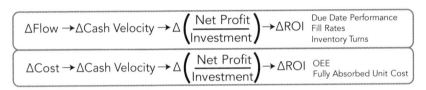

FIGURE 2-2   Flow versus cost perspectives and metrics

Figure 2-2 shows the critical difference in perspectives between a flow-centric approach and metrics versus a cost-centric approach and metrics. In most environments, both attempt to exist simultaneously, causing a direct breakdown in coherence as personnel struggle to find a constantly changing and tenuous balancing point where they minimize the damage for themselves and the company. This puts planning personnel into a constant state of conflict and stress.

Is Your Organization Ready for Flow?

These four prerequisites define what it means to think, communicate, and behave systemically—the only way to protect and promote flow. If an organization and its personnel have not embraced this thoughtware, then the flow of relevant information and materials will always be impeded to varying degrees. This directly leads to poorer ROI performance. Thus, before companies invest significant amounts of money, time, and energy into new hardware and software solutions, they should first consider investing in the proper thoughtware to gain visibility to what is truly relevant.

Here are some questions you can ask about your organization regarding visibility and coherence:

- Are people in your organization trained to think systemically? Without the capability to think and problem solve systemically, innovation and adaptation will be severely limited.
- Do they have a common systemic language and framework to think and work within? With different vernacular, nomenclature, and modes of operation for similar activity across different areas, it becomes difficult for one area to relate to another—there must be constant translation with critical information lost in translation.
- Do people in your organization understand the connections between departments, functions, resources, and people? Without understanding these connections, personnel cannot understand the total picture of flow. They may take actions that seem beneficial from a local consideration but are actually devastating to general company flow.
- Are people given enough visibility to understand the connections between departments, resources, and people? Without tools and processes to ensure visibility, personnel cannot keep evolving their knowledge of the system and identify system improvement opportunities.
- Are people discouraged from thinking and offering solutions outside of their operating area? If people are discouraged from thinking globally, you are guaranteeing that they will think locally.
- Do people understand how the different forms of variability affect the enterprise and *flow* through it? Without the ability to understand where and what variability to manage, people cannot take the necessary steps to protect flow in the system.

How does your organization score with these questions? If the answer is "poorly," it means that your organization lacks the proper thoughtware to successfully execute its mission to drive return on shareholder equity. If people do not understand the system or cannot see the entire system or are actively prevented from acting in the interest of the whole system, an organization is incapable of promoting and protecting flow.

Creating visibility to relevant information is no trivial task but it is the only path to sustainable organizational success. The more relevant information the organization has, the more immediate and enduring the success it will have—it is really that elementary.

Summary

If executives want to fulfill their mission to drive sustained return on shareholder equity, they must understand where to start. This starting point affects what they should and should not reinforce or reward. Driving sustainable ROI starts with visibility to relevant information. Producing that visibility has four prerequisites. The absence of any one of those prerequisites results in distortions that create variability and break down flow and coherence. This is exactly the case with the conventional supply chain management approach. This should not be an earth-shattering realization to anyone who has been in supply chain management for even a short period of time.

Companies need an operating framework with aligned metrics and a communication system that promotes visibility to the relevant information for the promotion and protection of flow to achieve and maintain coherence at the enterprise level. A blueprint and a step-by-step path are required to install this appropriate thoughtware. The Demand Driven Adaptive Enterprise model was designed to address this need and is described in the next chapter.

CHAPTER 3

The Demand Driven Adaptive Enterprise Model

The Demand Driven Adaptive Enterprise (DDAE) model is a complete management model enabling enterprises to sense market changes, adapt to complex and volatile environments, and develop market driven innovation strategies. It was designed specifically for the VUCA world. This model utilizes a constant system of feedback between three primary components: a Demand Driven Operating Model, Demand Driven Sales & Operations Planning, and Adaptive Sales & Operations Planning. A Demand Driven Adaptive Enterprise focuses on the protection and promotion of the flow of relevant information, materials and services to drive sustained return on equity performance.

The DDAE model spans the operational, tactical, and strategic ranges of an organization, allowing it to continuously and successfully adapt to the complex, changing, and volatile supply chains in existence today. It combines the fundamental principles of flow management with the emerging new science of complex adaptive systems (CAS). Successful businesses will need to work this way to survive and thrive in a VUCA world.

Figure 3-1 illustrates the essential components and basic mechanics of the DDAE model.[1] It is neither simply right-to-left nor left-to-right in nature nor outside in or inside out; it is all at the same time. The DDAE model is a bidirectional non-linear systems approach that drives adaptation through a cycle of configuration, feedback, and reconciliation across the three components.

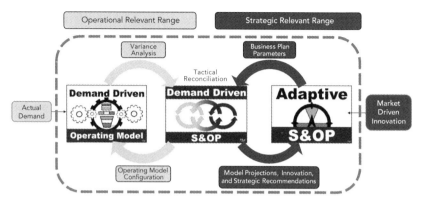

FIGURE 3-1   The Demand Driven Adaptive Enterprise (DDAE) Model

The Three Components of the DDAE Model

The DDAE Model has three distinct components designed to create a dynamic strategic, tactical, and operational ecosystem.

The Demand Driven Operating Model. A Demand Driven Operating Model (DDOM) is a supply order generation, operational scheduling, and execution model utilizing actual demand in combination with strategic decoupling and control points, and stock, time, and capacity buffers in order to create a predictable and agile system that promotes and protects the flow of relevant information and materials within the operational relevant range (hourly, daily and weekly). The DDOM's key parameters are set through the Demand Driven Sales & Operations Planning (DDS&OP) process to meet the stated business and market objectives while minimizing working capital and expedite-related expenses.

The Demand Driven Operating Model will be described in Chapter 4.

Demand Driven Sales and Operations Planning (DDS&OP). DDS&OP is a bidirectional tactical reconciliation hub in a Demand Driven Adaptive Enterprise (DDAE) Model between the strategic (annual, quarterly, and monthly) and operational (hourly, daily, and weekly) decision making relevant ranges. DDS&OP sets key parameters of a Demand Driven Operating Model (DDOM) based on the output of the Adaptive S&OP process (strategic information and requirements).

DDS&OP projects the DDOM performance based on strategic information and requirement inputs and various DDOM parameter settings. Additionally, DDS&OP uses variance analysis based on past DDOM performance against critical system metrics (reliability, stability, and velocity) to adapt the key parameters of the DDOM and/or recommend strategic changes to the business plan.

The six basic elements of DDS&OP will be described in Chapter 5.

Adaptive Sales and Operations Planning (Adaptive S&OP). Adaptive Sales and Operations Planning is the integrated business process that provides management the ability to strategically define, direct, and manage relevant information in the strategic relevant range across the enterprise. Market Driven Innovation is combined with the Operations Strategy, Go-to-Market Strategy, and Financial Strategy to create the strategic information and requirements scenario inputs for tactical reconciliation in addition to strategic projections to effectively create innovation and drive adaptation.

The seven elements of Adaptive S&OP will be described in Chapter 6.

What's Behind the DDAE Name?

Before we go into depth about the nature of the DDAE model, let's first understand the logic behind the name of the model. There are three elements of the name: the term "Demand Driven," the word "Adaptive," and the word "Enterprise."

Why Demand Driven?

The obvious answer can be seen in Figure 3-1. At each side of the DDAE model (the operational and strategic boundaries), there is a demand or market interface. On the left-hand side the model is combined with actual demand in order to drive operational activity. On the right-hand side, the model is combined with market-driven innovation to create desired future for the company. Market-driven innovation is unique to each company and rests largely in the vision of its executives and their ability to understand and articulate where and how the company will meet future opportunity in the market. With that said, a DDAE model allows the organization to more quickly respond to that vision and adapt when the market conditions change.

But there is much more to the term "Demand Driven" than that. The term was pioneered by PeopleSoft beginning in 2002. At the time, PeopleSoft was the second largest ERP company in the world. One of the leaders of that effort was Carol Ptak, Vice President of Manufacturing and Distribution Industries (later a co-founder of the Demand Driven Institute and a coauthor of this book). The term was then and still is defined as:

Sense changing customer demand, then adapt planning and pro-duction while pulling from suppliers—all in real time.

The problem was that at the time they really had no idea how to truly accomplish that vision. In 2003, PeopleSoft began to work with a services and technology company called Constraints Management Group, specializing in flow-based planning and manufacturing systems in order to help construct a "Demand Driven" approach. The Managing Partner of Constraints Management Group was Chad Smith (later a co-founder of the Demand Driven Institute and the other coauthor of this book).

In 2004, Oracle acquired PeopleSoft in a hostile takeover and the Demand Driven movement all but died. "Demand Driven" was then res-urrected in 2007 by Advanced Manufacturing Research (AMR). In 2010, AMR was acquired by Gartner, and Gartner used the term as part of what it called its "Demand Driven Value Network" approach. The problem was that the definition had been distorted to emphasize the objective for bet-ter forecasting while maintaining conventional supply order generation equations, including optimization and synchronization. This prompted Ptak and Smith to collaborate on a seminal work in 2011 that introduced a concept called Demand Driven Material Requirements Planning (DDMRP) to begin to set the record straight about what Demand Driven really is and how to practically implement it. But the articulation of DDMRP was only the first element of a much bigger body of knowledge.

Why Adaptive?

The model is designed specifically for a CAS and incorporates a framework for CAS characteristics to be better identified, focused, and utilized. As described in Chapter 1, CAS are constantly evolving through a cycle of emergence, feedback, and selection where:

- Emergence is a reconfiguration of the system triggered externally or internally.
- Feedback is a set of defined signals and triggers that are monitored by adaptive agents.
- Selection is decisions, actions and learning in response to the signals and triggers, which may or may not result in another reconfiguration.

The DDAE model uses two distinct adaptive loops or cycles to drive adaptation. These cycles are different because they operate in two different relevant ranges. The two cycles do, however, meet in the middle of the model with Demand Driven S&OP. Figure 3-2 depicts the two adaptive loops in the DDAE Model. Tactical Selection will be explained in Chapter 5 while Strategic Selection will be explained in Chapter 6.

Why Enterprise?

The DDAE framework is designed specifically for the enterprise and its multi-echelon subsystems to generate relevant information at all levels in order to mitigate variability so that flow can be promoted and protected now and in the future. This creates a resilient organization that is much better equipped to absorb disruption and adapt to the previously identified risks to resiliency.

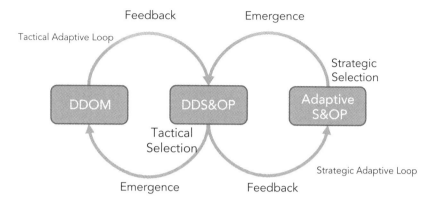

FIGURE 3-2 The adaptive loops of the DDAE Model

Summary

As mentioned before, DDMRP was only the first piece of the puzzle, the beginning of a journey to really enable the true vision of Demand Driven. A better supply order generation and management engine is simply not enough. Ptak and Smith founded the Demand Driven Institute and began to collaborate with experts around the world, rounding out a larger framework to enable that vision from top to bottom and bottom to top. They collaborated on additional books, further defining, expanding, and refining the topic. Educational programs of varying depths were developed to teach all levels of the organization this concept to facilitate true change. To validate that education, global certification standards were developed to ensure consistency in the concept. Finally, they worked with software companies to ensure compliance to these consistent standards. All of this has culminated in a true complete enterprise solution.

CHAPTER 4

The Demand Driven
Operating Model

The DDAE model incorporates a flow-based operating approach called the Demand Driven Operating Model (DDOM). Figure 4-1 shows the detailed schema of the Demand Driven Operating Model.

Before we explain what the DDOM is, let's first specifically state what it is not.

- DDOM does not mean "better forecasting." Forecast error is on the rise despite the proliferation of more advanced and more powerful algorithms. Supply chain complexity and volatility is increasing faster than we can compensate for it. Decoupling supply order generation from demand signals with known and significant error is a must as the penalties for guessing wrong are harsher than ever. The evidence for this was presented in Chapter 3. As such, a DDOM does not directly connect forecasts to actual supply order generation.
- DDOM does not mean "make to order everything." Whether we want to admit it or not we live in a "to-stock" world as supply chains have elongated and customer tolerance times have shrunk dramatically. The only choice for survival is that supply chains have to hold stock at some points. The customer is just not willing to wait! The DDOM relies on strategic stocking points at various

levels in the product structure and supply chains that are carefully monitored, adapted, and replenished. These stocking points determine the customer lead time and inventory investment.

- DDOM does not mean "inventory everywhere." A Demand Driven model uses strategically placed and managed points of inventory in order to dampen demand and supply variability simultaneously, compress lead times, and leverage common materials and components. Inventory is not simply spread everywhere—it is strategically positioned, appropriately sized, and carefully monitored for replenishment. This typically results in net reductions of system inventory between 25 to 50 percent while actually increasing service levels.

- DDOM does not mean "simple pull." Most MRP implementations represent a dramatic overcomplication; signals are constantly changing and conflicting. Conversely, most lean implementations represent an oversimplification as systemic visibility is extremely limited in the overall environment. Neither handles volatility well. For both methods to work effectively, the underlying assumption is that you are operating in a near perfect world— no volatility, no uncertainty, and no ambiguity. The DDOM was designed with all of this in mind. It successfully allows an environment to absorb variability, plan, and schedule at a multi-echelon enterprise level while pacing to actual demand so that a company can build what can and will be sold.

The DDOM is built around the same set of assumptions previously listed with the flow-based models in Chapter 2 in order to protect and promote flow for better ROI performance.

- Assets must be as closely synchronized to actual demand as possible.
- Variability cannot be eliminated but its impact on system flow can be effectively controlled with the right design.
- Focusing on precise planning and scheduling across all materials and resources is largely wasted effort.
- Some level of inventory is required at some stage in the supply chain.

- Batching decisions should be based on flow considerations instead of cost considerations.
- Some level of protective capacity must exist in any environment with multiple resources interacting.
- Inventory is an asset and should be treated as any other asset on the balance sheet.
- Capability must be established based on the expected future for the company.
- The objective of the operating model is to maximize contribution margin given the capacity capability within the relevant time period.

Figure 4-1 details the DDOM schema. The heart of the DDOM is the innovative method of strategic buffering, supply order generation, and execution known as Demand Driven Material Requirements Planning (DDMRP). DDMRP utilizes strategically determined decoupling point buffers to compress lead times and mitigate the distortion to relevant information (the transfer and amplification of demand signal distortion) up the supply chain and the distortion to relevant materials (supply continuity variability) down the supply chain.

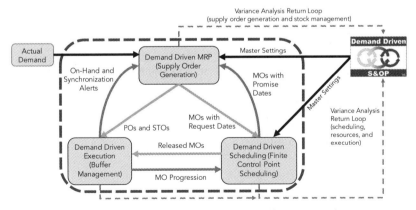

FIGURE 4-1   DDOM communication and settings schema

Then DMRP utilizes actual demand as part of its unique net flow equation in order to generate supply order signals. Manufacturing

Orders (MOs) with request dates are sent to Demand Driven Scheduling while Purchase Orders (POs) and Stock Transfer Orders (STOs) are sent directly to Demand Driven Execution. Real time status signals are then returned from each. From Demand Driven Scheduling comes manufacturing promise dates and from Demand Driven Execution comes on-hand and synchronization alerts.

Detailed resource scheduling is accomplished through Demand Driven Scheduling. Scheduling sequences Manufacturing Orders for execution. Execution returns order progression against the schedule. Demand Driven Execution is where stock, time, and capacity buffers are actively managed in relation to all open orders/scheduled activity.

One truly unique feature about the Demand Driven Operating Model is that there is no Master Production Schedule in use. The conventional MPS is replaced by the Master Settings input. Now it is possible for a company to build what can and will be sold because the operational capability range matches the strategic requirement. The settings and configurations for the entire DDOM (including DDMRP) are managed through the DDS&OP process. Closed-loop feedback is connected to DDS&OP from each of the three components of the DDOM, providing valuable variance analyses for future DDOM reconfiguration or adaptation.

The DDOM is designed around four basic elements.

DDOM Element #1: Strategic Decoupling Points

The Demand Driven Operating Model uses strategically placed decoupling points to compress lead times, shorten planning horizons, and dampen demand and supply variability simultaneously throughout the environment.

What Is Decoupling?

As discussed in Chapter 2, the bullwhip effect is the bidirectional transference, amplification, and accumulation of variability in both demand

and supply directions. The bullwhip effect directly impedes flow and the ability to drive, sustain, and improve return on investment performance. As the bullwhip effect is a bidirectional problem, the solution must also be one of a bidirectional nature.

The APICS dictionary defines decoupling as:

"Creating independence between supply and use of material. Commonly denotes providing inventory between operations so that fluctuations in the production rate of the supplying operation do not constrain production or use rates of the next operation." (Page 43)

Decoupling disconnects one entity from another. This isolates events that happen in one entity or portion of a system from impacting other entities or other portions of the system. Think of decoupling as a firewall isolating the events or environment on one side of the wall from the other side. Figure 4-2 shows the bidirectional impact of decoupling between two items (A & B). A brick wall is used to convey that the two items are decoupled; the wall isolates the variability (demand and supply) that occurs on one side from impacting the other side. Obviously, the height and/or thickness of this wall is dependent on the level of variability that must be stopped.

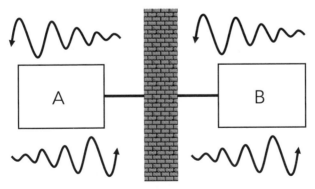

FIGURE 4-2   The impact of decoupling

Where to Decouple?

Decoupling promises a fundamental break from the hard-coded dependency calculations of MRP and APS where there is no decoupling. But the immediate question arises: where to decouple? Decoupling everywhere is an extreme position, one that has been proven to be problematic by the lean philosophy. Thus, the answer can only be to decouple somewhere. The points at which we choose to decouple are appropriately called "decoupling points." The APICS dictionary defines decoupling points as:

> "The locations in the product structure or distribution network where strategic inventory is placed to create independence between processes or entities. Selection of decoupling points is a strategic decision that determines customer lead times and inventory investment." (Page 43)

The transference and/or amplification of variability is dampened or stopped at the point where decoupling is imposed. There are two key observations to be made:

- At a single point, a bidirectional solution is created. Stopping the demand signal distortion and supply continuity variability does not require two different solutions; decoupling is just the opposite. Decoupling implies a relative simplicity that should be welcome in most supply chain and planning environments.
- The decoupling point does not eliminate variability at a discrete level; that is not its intention. Its intention is to mitigate the effect of variability on system flow. The bidirectional variability does begin to transfer and accumulate again in both directions after the decoupling point but the point is chosen to mitigate the impact and exacerbation of that variability on the flow of the system.

Figure 4-3 shows the decoupling point placement criteria for a Demand Driven Operating Model. There are six criteria. They must always be considered in combination since the choice of one can often impact the others.

Customer Tolerance Time	The time the typical customer is willing to wait before seeking an alternative source.
Market Potential Lead Time	This lead time will allow an increase of price or the capture of additional business either through existing or new customer channels.
Demand Visibility Horizon	The time frame in which we typically become aware of sales orders or actual dependent demand.
External Variability	• Demand Variability: The potential for swings and spikes in demand that could overwhelm resources (capacity, stock, cash, etc.). • Supply Variability: The potential for and severity of disruptions in sources of supply and/or specific suppliers.
Inventory Leverage and Flexibility	The places in the integrated bill of material (BOM) structure (matrix bill of material) or the distribution network that enable a company with the most available options as well as the best lead time compression to meet the business needs.
Critical Operation Protection	These types of operations include areas that have limited capacity or where quality can be compromised by disruptions or where variability tends to be accumulated and/or amplified.

FIGURE 4-3 Decoupling point placement criteria

Decoupling point buffer placement has significant implications for lead times. By decoupling supplying lead times from that certain point in a supply chain, lead times are instantly compressed to respond to the customer. This lead time compression has immediate service and inventory implications. Market opportunities can be exploited while working capital required in decoupling point buffers placed at higher levels in the product structure (end item or closer to complete) can be minimized.

Figure 4-4 shows a decoupling system design for a sample company. This is a mixed mode operation in which some end items are stocked while others are made to order. End items go through a series of resources that each have different processing times. The numbers within the circles are the number of minutes per piece at that resource. There is a decoupled intermediate component after the assembly operation. Additionally, the supplier inputs have all been decoupled.

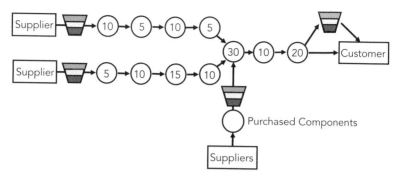

FIGURE 4-4   Sample decoupling point design

Decoupled Lead Time

The use of decoupling points also introduces the concept of decoupled lead time. Decoupled lead time is the longest unprotected sequence in a bill of material or distribution network. It is a qualified cumulative lead time between either two decoupling points, the last decoupling point and a customer, or the first decoupling point and a supplier.

Decoupled lead times are used as one of the primary components for sizing the levels of protection at decoupling points as well as exploring the implications of different decoupling placements (DDOM reconfigurations).

DDOM Element #2: Strategic Control Points

The Demand Driven Operating Model uses strategically placed control points for scheduling in addition to resource and order synchronization. Control points are places to transfer, impose, and amplify control through a system and/or within certain parts of system. Control points include

- **Pacing resources** determine the total system output potential. The slowest resource or consistently most loaded resource limits or defines the total system capacity for scheduling. These are commonly called a drum (consistent with the Theory of Constraints).

Drums are finitely scheduled based on actual demand or stock buffer replenishment orders. This finite schedule will then ultimately determine promise dates, material release times, and all other entry and exit point schedules.

- **Exit and entry points** are the boundaries of your effective control. Carefully controlling that entry and exit determines whether delays and gains are generated inside or outside your system.

- **Common points** are points where product structures or manufacturing routings either come together (converge) or deviate (diverge). One place can control many things.

- **Points that have notorious process instability** are good candidates because a control point provides focus and visibility to that resource and forces the organization to bring it under control or plan for, manage, and block the effect of its variability from being passed forward.

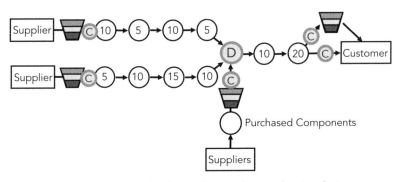

FIGURE 4-5   Sample decoupling and control point design

Figure 4-5 shows our sample design with control points including a drum added. Control points that are entry and exit points have been labeled with a "C" inside of a circle. The control point acting as a drum is labeled with a "D" inside of the circle. In this case the resource that took the longest per piece and was also an integration/common point in the routing was selected as a drum. Alternative demand driven scheduling for this drum position could be a heijunka board or rhythm wheel schedule. Whether the schedule simply minimizes the number of early and late orders or sequences the orders in a specific manner to minimize the total set up time at the drum, the key point is that this is the relevant resource schedule that drives the schedule for initial release and provides information for order promising.

DDOM Element #3: Dynamic Buffering

The Demand Driven Operating Model protects its decoupling and control points through dynamic stock, time, and capacity buffers.

Stock Buffers

For the decoupling points to maintain their decoupling integrity, there must be a level of protection that absorbs demand and supply variability at the same time. This level of protection is a concept called "decoupling inventory." The APICS dictionary defines decoupling inventory as:

> "An amount of inventory kept between entities in a manufacturing or distribution network to create independence between processes or entities. The objective of decoupling inventory is to disconnect the rate of use from the rate of supply of the item." (Page 43)

In the Demand Driven Operating Model, decoupling point inventory is also commonly referred to as "decoupling point buffers" or "stock buffers." Decoupling point buffers are quantities of inventory or stock that are designed to decouple demand from supply. Decoupling point buffers are amounts of inventory that will provide reliable availability for stock consumption while at the same time allow for the aggregation of demand orders, creating a more stable and efficient supply signal to suppliers of that stock.

Sizing Decoupling Point Buffers

Decoupling point buffers are sized through a pragmatic, proven, and transparent process. They incorporate a three-color zonal schema

(Green, Yellow, Red). Each zone serves a specific purpose in the way it protects and helps to manage the decoupling point. Figure 4-6 shows the purpose of each zone of the buffer.

Green	Determines order size and frequency
Yellow	The heart of demand coverage
Red	The safety in the buffer

FIGURE 4-6   The purpose of each decoupling point buffer zone

The zonal values and total buffer value is determined through a combination of two elements. First, part positions chosen for decoupling are sorted or grouped by like attributes called buffer profiles. These attributes are:

- The type of part (manufactured, intermediate, purchased, or distributed)
- Lead time category (long, medium, short)
- Variability category (high, medium, low)

Next, the individual part traits are combined with these attributes to create a unique zonal and total buffer values. Figure 4-7 depicts this basic formula and the individual part attributes of average daily usage (ADU), decoupled lead time, minimum order quantity (MOQ), and location (if applicable).

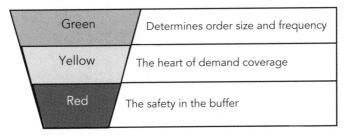

Part Trait	Buffer Profile Assignment		Buffer and Zone Levels
Average Daily Usage (ADU)		Part Type	
Decoupled Lead Time	X		=
Minimum Order Quantity (MOQ)		Variability Factor	
Location (Distributed parts only)		Lead Time Factor	

FIGURE 4-7   Decoupling point buffer sizing factors

Figure 4-8 is an example of the full calculations for a buffer. This part is assigned to a buffer profile called MML. That means it is a manufactured item, with medium lead time and low variability assignments. It has an average daily usage of 10 parts. This average daily usage is most often calculated based on part demand history over a user-defined horizon. There is a minimum order quantity of 50, which is used as a potential sizing input to the green zone only. There is also an imposed order cycle of seven days. This means that the part must be ordered every seven days for scheduling reasons. Like minimum order quantity, order cycle is only a potential determinant of the green zone value.

Green Zone Calculation
Step 1: Calculate the Green Zone using the Lead Time Factor. ADU (10 per day) × DLT (12 days) × Lead Time Factor (.5) = 60

Step 2: Compare the calculated green zone in step one against the minimum order quantity and imposed order cycle factor and take the greater. The calculated green zone of 60 is larger than the minimum order quantity of 50. The imposed order cycle of seven days, however, produces the largest green zone at 70 (ADU × Order Cycle). This part's buffer green zone will be set at 70 pieces.

Yellow Zone Calculation
The yellow zone is extremely easy to calculate. ADU (10 per day) × DLT (12 days) = 120

The Red Zone
Step 1: Calculate the Red Base using the Lead Time Factor. ADU (10 per day) × DLT (12 days) × Lead Time Factor (.5) = 60

Step 2: Calculate Red Safety using the Variability Factor. 60 (Red Base) × .33 = 20

Step 3: Calculate the total Red Zone by adding Red Base plus Red Safety. Red Base (60) + Red Safety (20) = Total Red Zone (80)

The total buffer value called "Top of Green" is 270 (70 + 120 + 80). It is important to understand that this is only a plan-to level; average on-hand will be much lower. Average on-hand is calculated as half the green zone + the total red zone. In our example, that would yield an average on-hand position of 115 units (80 + 35). This is called the on-hand target position. The average inventory range is defined as the top of the red zone value to the top of the red zone plus the green zone values. In our example (depicted below) this would be 80 (top of red zone) to 150 (top of red plus green). The on-hand target position and range is a primary performance metric foundation for DDOM analytics to enable model adaptation.

Example Part Buffer Calculation				
Average Daily Usage	10	Green Zone	70	
Buffer Profile	M, M (.5), L (.33)		LT Factor: 60 (DLT (12) x ADU (10) x Lead Time Factor (.5))	
MOQ	50		Minimum Order Quantity: 50	
Imposed or Desired Order Cycle (DOC)	7 days		Order Cycle: 70 (7(OC) x 10(ADU))	
Decoupled Lead Time (DLT)	12 days	Yellow Zone	120 (12(DLT) x 10(ADU))	
		Red Zone	80 (Red Base (60) + Red Safety (20))	
			Red Base: 60 (DLT (12)x ADU (10) x Lead Time Factor (.5))	
			Red Safety: 20 (Red Base (60) x Variability Factor (.33))	

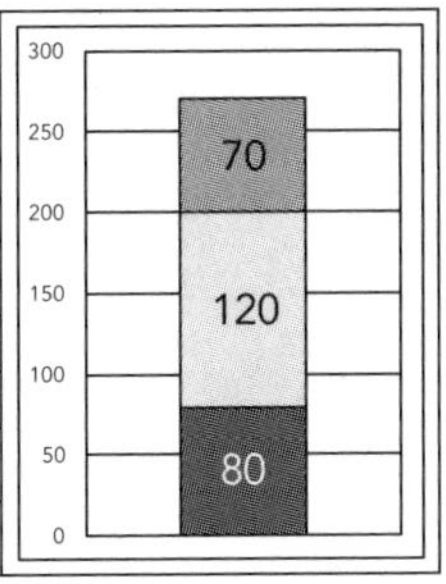

(Bar chart, scale 0–300, showing: Green 70, Yellow 120, Red 80)

FIGURE 4-8   Example of calculated decoupling point buffer

For a more detailed look at how these buffers are calculated, please refer to Chapter 7 of *Demand Driven Material Requirements Planning Version 2* (Ptak and Smith, Industrial Press, 2018).

Adjusting Decoupling Point Buffers

Since today's supply chains are incredibly dynamic, the initially determined buffer and zone levels must adjust and adapt to changing conditions. Understanding the equations to set the buffer zones means that the factors that can change a part's buffer size over the course of time are also understood. These changes can come from any part attribute change and/or buffer profile changes.

There are two basic forms of buffer adjustments: those that are implemented as a result of what has happened and those that are implemented based on what we think will or plan to make happen. Figure 4-9 shows the most basic form of buffer adjustment, called a recalculated adjustment. A recalculated adjustment is simply an automated adjustment to a buffer based on changing attributes or profiles. In this case the only attribute that is changing is the calculated average daily usage (ADU). The buffer and zone values relate to the left hand Y axis while the changing ADU relates to the right hand axis.

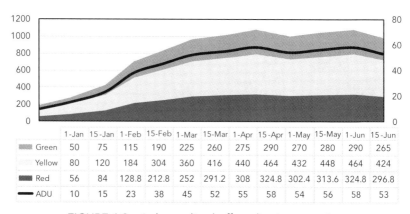

	1-Jan	15-Jan	1-Feb	15-Feb	1-Mar	15-Mar	1-Apr	15-Apr	1-May	15-May	1-Jun	15-Jun
Green	50	75	115	190	225	260	275	290	270	280	290	265
Yellow	80	120	184	304	360	416	440	464	432	448	464	424
Red	56	84	128.8	212.8	252	291.2	308	324.8	302.4	313.6	324.8	296.8
ADU	10	15	23	38	45	52	55	58	54	56	58	53

FIGURE 4-9   A decoupling buffer adjusting over time

Buffers also can be manipulated through planned adjustments. Planned adjustments are based on certain strategic, historical, or business intelligence factors as identified in tactical planning. These planned adjustments are manipulations to the buffer equation that affect target inventory planning positions by raising or lowering buffer levels and their corresponding zones at certain points in time. These manipulations tend to be confined to demand input manipulations, zonal manipulations, or lead time manipulations.

Figure 4-10 shows a buffer that has been adjusted through a demand adjustment factor (DAF). The original ADU is displayed on the top row of the data table. The second row displays the value of the DAF being applied. The third row then shows the adjusted or factored ADU. The DAF is applied to the ADU in order to create a temporary increase in buffer levels and is commonly applied for seasonality or large promotions.

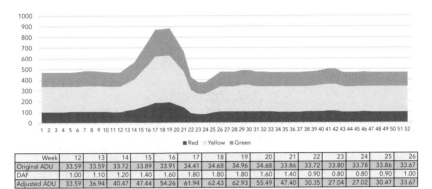

Week	12	13	14	15	16	17	18	19	20	21	22	23	24	25	26
Original ADU	33.59	33.59	33.72	33.89	33.91	34.41	34.68	34.96	34.68	33.86	33.72	33.80	33.78	33.86	33.67
DAF	1.00	1.10	1.20	1.40	1.60	1.80	1.80	1.80	1.60	1.40	0.90	0.80	0.80	0.90	1.00
Adjusted ADU	33.59	36.94	40.47	47.44	54.26	61.94	62.43	62.93	55.49	47.40	30.35	27.04	27.02	30.47	33.67

FIGURE 4-10   Buffer with demand adjustment factor (DAF) applied

It should also be noted that this is *not* connecting forecast directly to supply order generation as convention does. At the most this is connecting anticipated or forecasted demand to the level of protection that may be required within a period of time. The adjusted buffer may or may not drive a supply order according to the new adjusted level. Supply order release depends entirely on the planning mechanism of the buffer called the net flow equation. The DAF is typically adjusted based on the availability of capacity to balance the expected factory load. The actual load is driven by actual replenishment orders.

Replenishing Decoupling Point Buffers

First and foremost, a critical point of understanding about decoupling points is that they are *never* netted to zero. Netting a decoupling point to zero means that it ceases to be a decoupling point. Thus, a level of decoupling point inventory must be maintained in order to guarantee a decoupling point's effectiveness.

In the DDOM, decoupling point buffers use a unique equation called the net flow equation to drive replenishment order release. The demand element of the net flow equation has no forecasted orders; only qualified sales orders within the immediate time horizon are considered. A critical difference is in the quality of the demand signal. A sales order is highly accurate—it is an explicit statement of what will be consumed while planned orders based on forecast are highly inaccurate.

The net flow equation is simple:

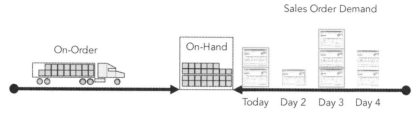

FIGURE 4-11   Illustrating the elements of the net flow equation

On-Hand: The quantity of stock physically available. In Figure 4-11 this is represented by the large box labeled "On-Hand" in the middle of the graphic. The smaller boxes represent the actual quantity on-hand and available for use.

On-Order: The quantity of stock that has been ordered but not received. In Figure 4-11 this is represented by the truck on the left hand side of the graphic that is heading toward the on-hand position. The smaller boxes in the trailer represent the amount on order. This could be a single incoming order or several incoming orders. The on-order quantity is the total quantity that has been ordered but not received, irrespective of timing.

Qualified Sales Order Demand: The sum of sales orders past due, due today, and qualified spikes. In Figure 4-11 this is represented by the orders that are highlighted "Today" and in "Day 3." There is no past due amount represented in this figure. There are two sales orders due today and another three sales orders in Day 3 that have combined to create a qualified spike for that day. These two days of qualified demand are added together to get the total amount of qualified demand for today's computation of the net flow equation. Order spikes are qualified through the combined use of an order spike horizon in future daily buckets and an order spike threshold.

When the net flow position is below the bottom of the green zone, a supply order is issued to restore the net flow position to the top of green. Figure 4-12 shows how the net flow equation works. This buffered item has a green zone of 120, a yellow zone of 240, and a red zone of 95.

That is a top of green value of 455. Today this item's net flow position is 325 (105 (on-hand) + 240 (on-order) – 20 (qualified demand). That puts the net flow position into the yellow zone, which will result in a recommended supply order for a quantity equal to the amount to restore the net flow position to the top of green. That quantity is 130 units.

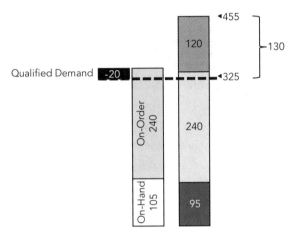

FIGURE 4-12   The net flow equation

A key aspect of using the net flow equation is the visibility that planning personnel gain to relative requirements priority. This relative priority distinction is a crucial differentiator between the conventional MRP planning alerts and action messages and the highly visible and focused DDMRP approach. Conventional MRP is a binary system. You are either OK or not OK with regard to each part. There is little sense of how parts compare to each other—you need to either act or not act having no idea of how OK or not OK each part that is displaying an action message is.

Under the DDMRP approach, planners and buyers can quickly judge the relative priority without massive amounts of additional analysis and data queries. This is accomplished through the use of planning priority. Planning priority incorporates two aspects: color and a percentage value. The color is established by identifying the zone color of the net flow position. The percentage is computed by dividing the net flow position by the top of green value.

Part#	Planning Priority	On-Hand	On-Order	Qualified Demand	Net Flow Position	Order Recommendation	Request Date	Top RED	Top YELLOW	Top GREEN	Lead Time
406P	RED 19.8%	401	506	263	644	2606	4-Aug	750	2750	3250	20
403P	YELLOW 43.4%	1412	981	412	1981	2579	23-Jul	1200	3600	4560	8
402P	YELLOW 69.0%	601	753	112	1242	558	24-Jul	540	1440	1800	9
405P	YELLOW 74.0%	3400	4251	581	7070	2486	24-Jul	1756	7606	9556	9
401P	YELLOW 75.1%	2652	6233	712	8173	2715	25-Jul	2438	8938	10888	10
404P	GREEN 97.6%	1951	1560	291	3220	0		1050	2550	3300	6

Today's Date: 15-July

FIGURE 4-13   Relative priority in a DDMRP planning screen

Figure 4-13 illustrates a DDMRP planning screen showing both color and percentage. The sequence is determined first by color and then by the planning priority column percentage. The lower the percentage, the higher the planning priority. This sequencing now provides relative priority across multiple parts calling for resupply. This is crucial when limitations or constraints are present in an environment. When dollars, time, space, and resource capacity are at a premium, it is extremely advantageous to be able to quickly focus on the highest priority requirements and not spend critical resources on items that can wait.

Time Buffers

Demand Driven time buffers are planned amounts of additional time inserted in the schedule in order to cushion a control point schedule from disruption. Time buffers create a small queue of work waiting to be processed by or passed through a control point at a scheduled time. Time buffers are sized based on the reliability of the sequence of resources feeding the control point. The less reliable or more variable that string is, the larger the time buffer requirement.

Figure 4-14 shows how a time buffer conceptually works. The time buffer is the green, yellow, and red strip in front of the control point icon of a radio tower. The size of this time buffer is nine hours. That means that work will be scheduled to arrive nine hours in advance of the control point scheduled start.

The buffer is divided equally into three color zones, each of a three-hour duration that will determine priority. Upstream processes deliver work orders into the buffer. Those arrivals are subject to variability sym-

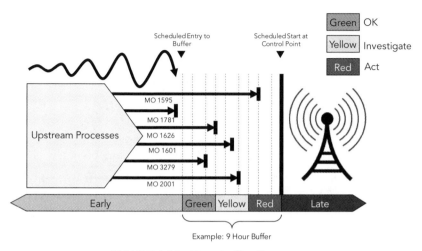

FIGURE 4-14 Time buffer illustration

bolized by the wavy line moving from left to right. When a work order fails to arrive in the buffer at the scheduled entry, it first penetrates the green zone of the buffer. Green zone penetrations are expected and normal—the buffer is expected to be used. For example, Manufacturing Order (MO) 3279 is now two hours into the green zone.

After three hours of an order's continued absence to the buffer, its status changes to the yellow zone. A yellow zone entry is an escalation to the relative urgency associated with that order. At this point, appropriate personnel should be investigating the absence. For example, MO 2001 is two hours into the yellow zone and has not yet arrived to the buffer.

After six hours of an order's continued absence in the buffer, its status progresses to the red zone. The red zone is further escalation in urgency, typically calling for an immediate action. Action could include expediting and/or resequencing planned work at the control point. MO 1595 is one hour into the red zone. It should be noted that order arrival in the red zone does not disrupt the established control point schedule; it means that it almost disrupted the schedule.

If an order has not arrived in the buffer by the time it is scheduled, then it is late; scheduled flow has been disrupted. Monitoring the frequency of red and late zone entries is crucial in understanding how to focus improvement activities in upstream processes. By eliminating red and late zone entries to time buffers, companies can effectively and safely shrink the amount of time incorporated in their time buffers. This has a direct effect on inventory levels and lead time.

Capacity Buffers

Capacity buffers protect control points and decoupling points by giving resources in the preceding work flow the available capacity to "catch up" with delay variability. Thus, a capacity buffer is protective capacity that provides agility and flexibility. The size of the capacity buffer allows both stock and time buffers to be reduced.

This protective capacity is available each day in every resource that is not capacity constrained. Capacity buffers are not to be used to improve unit cost or to maximize a particular resource's utilization. In fact, the entire notion of a capacity buffer is the antithesis of conventional costing policies. Capacity buffers require that a resource maintain a bank of capacity that goes unused so that the schedule at control points can be better maintained.

Figure 4-15 illustrates the concept of a capacity buffer. The resource's total planned capacity is defined by the top of the red zone. As scheduled load for each day approaches the top of the red zone, the resource becomes more and more likely to become constrained and potentially disrupt flow. The more constrained it becomes, the more likely that deep penetrations into the time buffers will begin to occur. This detailed capacity loading view (commonly referred to as resource loading charts) becomes crucial in identifying and properly managing capacity buffers on a day-to-day basis.

Exploring ways to minimize the investment and expense associated with this unused capacity is absolutely valid. What is *not* valid is encouraging a resource to misuse its spare capacity to improve unit cost or resource efficiencies by running unnecessary production; that is purely distortive and wastes the company's resources. When that happens, market responsiveness goes down and the stock and time buffers are jeopardized, forcing them to be increased to compensate. This will result in increases in lead times and inventory levels. This is precisely the wrong way to "utilize" a company's assets.

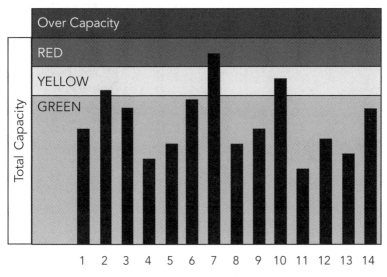

FIGURE 4-15   Capacity buffer illustration

Figure 4-16 illustrates a complete DDOM with all buffers (stock, time, and capacity) depicted.

Figure 4-17 shows our sample company DDOM design with the appropriate order signals. The time and capacity buffers have been removed for graphical simplicity. Sales orders are relayed for inclusion in the net flow equation for the end item buffer. Sales orders for make to order items are relayed along with required end item replenishment manufacturing orders for inclusion in the intermediate item buffer net flow equation. Replenishment manufacturing orders for the intermediate items are then sequenced at the drum based on the resource's finite capacity as discussed previously. That sequence then determines the corresponding material release schedules. Finally, purchased item stock buffers launch purchase orders based on their respective net flow positions.

Readers wishing to learn more about the details of the Demand Driven Operating Model should read *Demand Driven Performance—Using Smart Metrics* (Smith and Smith, McGraw-Hill, 2014).

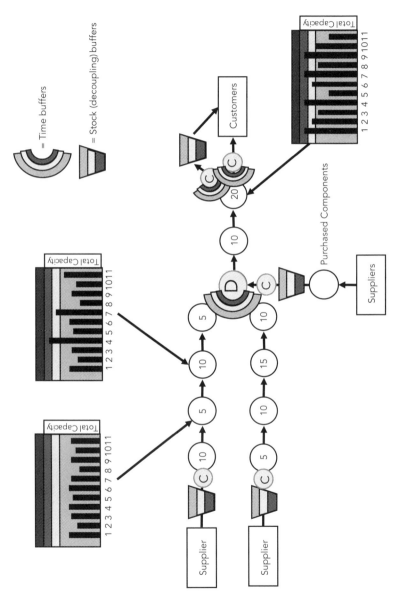

FIGURE 4-16 Sample DDOM design with all buffers

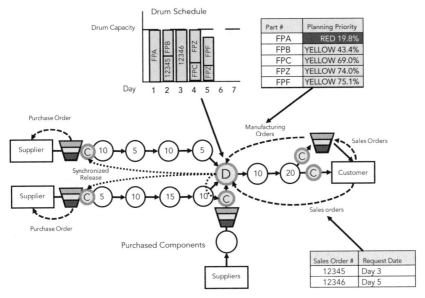

FIGURE 4-17   Sample DDOM design with order signals

DDOM Element #4: Pacing to Actual Demand

As explained previously, the Demand Driven Operating Model uses only actual demand rather than forecasted demand for supply order genera-tion. There are no planned orders and no Master Production Schedule used in the DDOM. Actual demand is the most relevant and undistorted demand signal available to companies; it is an explicit statement of need. When actual demand drives higher level decoupling point replenish-ment, then subsequent exploded demand to the next tier of decoupling points is connected to that actual demand.

How can we understand the practical implementation and impact of this on the order planning and requirements generation mechanism? As discussed earlier, decoupling points are strategic in nature and carefully selected. We don't simply decouple everywhere. That implies that there are still dependent points and/or sequences at play.

Blending the use of decoupling with the notion of managing and accounting for dependencies is accomplished through the use of a decoupled explosion. A decoupled explosion is the cessation of a dependent requirements explosion at any decoupling point. The term itself seems to be an oxymoron. It literally means independent dependence. Yet that is exactly what is occurring with decoupled explosion.

When a supply order is generated at a higher level, decoupling stops the explosion of the bill of material at the decoupling points placed at lower levels. The explosion can be stopped without risk because that decoupling point is buffered with decoupling inventory. The explosion then restarts only when the decoupled position (through an independent net flow calculation) determines that it needs resupply.

Figure 4-18 shows the concept of decoupled explosion in a sample product structure. In this example, high-level demand from FPA explodes through the product structure driving dependent demand requirements to SAA, SAB, and the supplier for PPA. The bold X denotes the decoupling points. The explosion then independently restarts at the appropriate time and for the appropriate quantity (according to its net flow position) at SAA and SAB. SAB's explosion will drive through to the supplier for PPB and PPC. SAA's explosion will drive to PPD, PPE, PPF, and PPG, where it will cease. These decoupled purchased items will then independently call for resupply at the appropriate time and for the appropriate quantity based on their independently determined net flow positions.

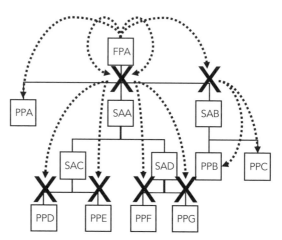

FIGURE 4-18   Illustrating decoupled explosion

This type of explosion is obviously different than conventional MRP where high-level demand over a planning horizon is typically driven all the way through to the purchased component/material level. There are some exceptions to this rule in MRP but they are simply that—exceptions.

While there are obvious differences, there are also similarities between a conventional MRP explosion and a decoupled explosion. There is independence at the decoupling points but between decoupling points there is dependence. That dependence between decoupling points is no different than conventional MRP. Figure 4-19 displays the same product structure from Figure 4-18 comparing a conventional MRP explosion and a decoupled MRP explosion. There are areas in which the explosion behaves in *exactly* the same manner. One of these sets of connections is highlighted with a dashed box (the connection SAA to SAC and SAD).

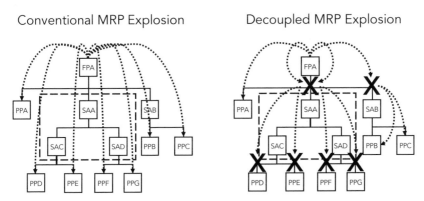

FIGURE 4-19   Conventional MRP explosion versus a decoupled explosion

The Operational Relevant Range

Now that we have described the components of the DDOM, we are in a position to better identify specifically what the operational relevant range is. The most obvious answer is that the operational relevant range is typically defined at the hourly, daily, or weekly consolidation level. But the DDOM provides a very specific answer: the decoupled lead time.

As mentioned before, decoupled lead time is the longest unbuffered sequence in time between:

- The supplier and a decoupling point (most often a purchase part or material buffer).
- Two decoupling points within a product structure and or distribution network.
- The last decoupling point and a customer delivery.

Therefore the decoupled lead time relative to the resource involvement or part number is the only reasonable answer for resources. Decoupling points are designed to isolate one side from the other. Thus, mixing the relevant ranges of one side with the other would be not only distortive but unfair to operational personnel. Worse yet, why would we ask people on a manufacturing floor to execute to metrics that consider factors well outside their relevant range, such as the fixed cost factors included in fully absorbed unit cost? Do we really believe that encouraging an operator to run a larger batch will actually affect our fixed cost structure? No? Then why do it? Why not provide metrics that focus operations on flow in the here and now working with the defined flow-based model and with considerations that are under their direct control?

Does that mean that what is happening on one side of a decoupling point should not impact what is happening on the other side? Of course not, but that consideration is at a higher level of system monitoring and tactical planning.

Operational Flow-Based Metrics

Flow-based operational metrics must emphasize system reliability, stability, and velocity to determine relevant information and materials for personnel in the operational relevant range. Remember that the operational relevant range is the short-range future—the here and now. As such, any metrics for the relevant range must reinforce what operational personnel need to be doing at any particular time to enhance short-term flow. Additionally, any metrics that conflict with those flow-based metrics must be removed or conflicts will be created that will inhibit flow and risk coherence to the DDOM.

Figure 4-20 displays the three basic metric objectives for the DDOM. They are not specific metrics, they are objectives that metrics should reinforce. The specific metrics will be tailored for each DDOM and can even change as the DDOM changes and evolves. Additionally, in Figure 4-20 the message that the specific metric should convey to those being measured is included. Measurements directly lead to and reinforce behavior. If a metric drives a distorted message, we can be assured that the behavior will also be distorted. Remember, the right metrics should neither be a source nor an amplifier to system variability.

Metric Objectives	The Message Behind the Objective
Operational Reliability	Execute to the model, plan, schedule, and market expectation
Operational Stability	Pass on as little variation as possible
Operational Speed/Velocity	Pass the right work on as fast as possible

FIGURE 4-20   The operational metric objectives

Operational reliability is fundamentally about aligning metrics to meet the expectations of the DDOM design. Does our behavior relative to the operation of the model reflect the design intentions and assumptions? Example metrics include

- **Net Flow and On-Hand Stock Status of Strategic Buffers.** A DDOM cannot produce a reliable output if the integrity of the decoupling points cannot be maintained now and into the short-term future.
- **Order Acceptance and Order Launch Timeliness.** A DDOM will be more reliable if relevant information is conveyed in a timely fashion. Supply order signals should be approved according to the model without significant delay. Order release signals to manufacturing should occur when scheduled in order to ensure

synchronization to critical post-release schedules and provide relevant information for shop floor execution.

- **Control Point Schedule Maintenance.** A DDOM's reliability will be challenged when control point schedules are not maintained—particularly at the last control point before the customer.

Operational stability is about actively controlling the transference and amplification of variability. To control variability does not mean to eliminate it—that is a fool's quest. Controlling variability means that we are mitigating its effect on total system flow. Operational stability is all about monitoring and properly acting on stock, time, and capacity buffer statuses. Example metrics include

- Measuring the current stock buffer status and/or material synchronization alerts of purchased and intermediate components/materials. A DDOM cannot have a stabilized schedule if components and materials are missing.
- Measuring the number of late arrivals to stock and time buffers in addition to monitoring control point schedules. The stability of the DDOM is challenged when materials, components, and end items are not available as planned.

Operational velocity is about encouraging speed while minimizing variability by focusing on order progression in relation to critical scheduled activity through the use of "Flow Exception Reporting." This type of exception reporting tracks variances in actual flow to scheduled flow with regard to order release schedules, rate of progress to the next critical scheduled activity, and critical scheduled activity (control point schedules). Example metrics include

- Measuring the number of late arrivals to stock and time buffers and control point schedules.
- Measuring actual work order progression rate to planned rate. Work orders particularly on long routings that lag may portend a buffer challenge in the near future.

We should stress that these metrics aren't analytical in nature. They are not measuring the performance capability of these objectives over time; that is the job of DDS&OP. These operational metrics are simply designed to encourage people to behave in line with the model in the here and now or immediate future (the operational relevant range), much like a stop light. If the light is red, you stop and wait. If you choose to keep going, a camera takes a picture of your license plate and you get a ticket in the mail. Why? Because while it advances your individual flow, you are risking overall system flow through the intersection. People may have to wait for you to clear the intersection or, worse yet, a collision can occur that will shut down the entire intersection for a period of time.

These operational metric objectives obviously dovetail together and reinforce each other. It should be apparent that there are overlapping specific metrics; one particular metric may be able to contribute to more than one objective. This fact is crucial in understanding the purpose of metrics in a flow-based system. Metrics should neither be a source of nor an amplification of variability. Metrics should help provide a clear course of action across a wide array of specific scenarios in order to protect and promote flow.

Summary

This chapter is in no way sufficient to understand the details in breadth and depth of the Demand Driven Operating Model. There are several resources available including books and courses that can provide critical and necessary personnel with those details.

What should be apparent is the emphasis of the DDOM on visual, intuitive, and real-time signals for operational personnel and/or adaptive agents to facilitate flow. The DDOM is a high-level management by exception method. You operate to the schedule until a signal cues operations to make a change based on a larger threat to the overall short-term reliability, stability, or velocity of the model.

But what ensures that the DDOM is designed sensibly and adapts based on its past performance and what the future will most likely bring? Without adaptation, the system's overall reliability, stability, and velocity will break down even if operational personnel are doing as they should. This evolution and adaptation is the responsibility of the tactical reconciliation component of the DDAE Model: Demand Driven Sales & Operations Planning.

CHAPTER 5

Demand Driven Sales and Operations Planning

Now our attention turns to the tactical reconciliation hub of the Demand Driven Adaptive Enterprise Model. While the conventional planning approach recognizes each of the relevant three ranges identified in Chapter 2, there has been a huge disconnect—a missing link—in the effectiveness of the Sales and Operations planning process to connect from intended strategy to day-to-day operations. As a reminder, Figure 2-1 is below.

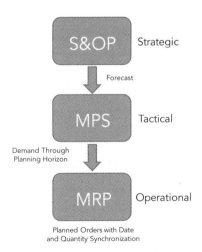

FIGURE 2-1   The conventional approach

Traditionally, S&OP has attempted to manage the portfolio and new activities, demand, and supply plans and reconcile them with the business plan through a management business review. Successful S&OP implementations have a robust process where management considers marketplace information to communicate a feasible business plan to operations that meets the financial business requirements. However, that connection traditionally has been through a Master Production Schedule (a statement of what can and will be built—one number supported by expected production by part number by specific date) that feeds the formal planning system. These single number schedules driving supply order generation and operational activity will be precisely wrong and will result in massive amounts of corrective actions as customers don't demand what is in that plan. This will require adjustments to the outcome in order to be even close to approximately right.

An effective S&OP plan is not and never has been a single number to run the business; S&OP is not about balancing supply and demand. Instead, truly effective S&OP is about ranges—an expected number with a pessimistic lower range and an optimistic upper range. This range represents the intended strategic direction with allowable boundaries from the executive team. However, traditional formal planning cannot calculate from a range. The MRP system needs a demand plan that is precise in quantity and timing. This is the chasm that must be crossed from the strategic range to the operational range, a role traditionally filled by the MPS. Unfortunately, this is like trying to cross the Grand Canyon on a single wire. If you can balance precisely, there is no wind, and nobody disrupts that wire, you may get to the other side rather than falling to your death. The VUCA world we must now manage is akin to someone shaking that wire and having gale force winds coming through the canyon at the same time.

Thus, a new bridge is required to cross this canyon—Demand Driven Sales and Operations Planning (DDS&OP). DDS&OP is a bidirectional tactical reconciliation hub in the Demand Driven Adaptive Enterprise (DDAE) Model between the strategic and operational relevant decision making ranges. DDS&OP sets key parameters of a Demand Driven

Operating Model (DDOM) based on the strategic information and capabilty requirements output of the Adaptive S&OP process. DDS&OP also projects the DDOM performance based on these strategic information and capability requirements in addition to various DDOM parameter settings. DDS&OP uses variance analysis based on past DDOM performance against critical relevant metrics to adapt the key parameters of the DDOM and/or to recommend strategic changes to the business.

DDS&OP is typically performed by a DDS&OP Team. The DDS&OP Team members are the adaptive agents managing the tactical adaptive cycle that is the bidirectional connection between strategy and operations. The composition and size of this team is relative to the size and complexity of the organization. In major corporations it would be a dedicated team with the following characteristics:

- Members that have a strong grasp of operations management and demand driven methods
- Members with an in-depth knowledge of the company's DDOM (both design and demonstrated capability)
- Members with good communication skills
- Representation or liason from Finance
- Representation or liason from Sales and Marketing
- Representation or liason from Engineering can be helpful

It may be beneficial to have the DDS&OP team report directly to the CEO, CFO, or COO of the organization to help remove any potential conflicts of interest between specific business groups. DDS&OP has six basic elements, all of which relate to the tactical flow-based metric objectives.

The Tactical Relevant Range

This brings us to a critical question: what exactly is the duration of the tactical relevant range? To define this range we must understand the basic purpose of the tactical adaptive cycle. The tactical adaptive cycle is about reconfiguring and adapting the Demand Driven Operating Model. To do that, a DDS&OP team must constantly consider these three basic questions:

- How has the model performed in the recent past?
- How is the model currently performing? Are there any significant conflicts currently inhibiting flow across the resource base or are there opportunities to increase flow in the near term?
- How is the model expected to perform given key current information about the future?

With these three questions in mind, a uniquely defined relevant range is revealed; it is one that incorporates the near-term past and the near-term future. We also must be reasonable about the speed with which change can effectively occur in any supply chain and manufacturing entity. The defined horizon that MRP systems have relied on since their inception is called cumulative lead time.

Cumulative lead time is the longest sequence of dependent activities within a given product structure. It is calculated by adding up the longest path of assembled, manufactured, and purchased components in a bill of material. As discussed in Chapter 4, the placement of strategic decoupling points breaks up cumulative lead time into a series of decoupled lead time horizons, the longest of which defines the operational relevant range for specific products and resources.

Average cumulative lead times will differ dramatically between industries. In aerospace, a typical cumulative lead time could be 12 months, while in consumer products it could be 45 days. Theoretically (because MRP was designed to net to zero), the cumulative lead time defines the cycle time of a particular product structure. This means that any significant changes relative to product structure must typically take at least one cumulative lead time to show up at the end item or customer level.

Changes, of course, can be made at any time at the resource level but there are some basic things to consider regarding changes to capacity/resources:

- Shorter-term resource decisions can be made that will affect flow such as overtime or outsourcing. These are variable expenses and do not affect the fixed expense base of the business.
- Longer-term resource decisions can be made that do affect the fixed expense base. A shift can be added, a new machine center can be purchased, or new plant can be built.
- In most environments the longer-term resource decisions take longer to implement than the cumulative lead time.

With all these considerations in mind, the Demand Driven Adaptive Enterprise Model defines the tactical relevant range as one cumulative lead time in the past plus one cumulative lead time in the future. This tactical range will define the scope of DDS&OP activities.

Figure 5-1 displays the DDS&OP schema. To the left, the bidirectional connection to the DDOM can be observed. Master settings configure the DDOM while a feedback loop in the form of variance analysis returns information to DDS&OP. To the right, the bidirectional connection to Adaptive S&OP can be observed. The validated business plan is provided by the Adaptive S&OP process. This process will be described in the next chapter. That plan has an estimated demand range, capabilities, and future performance targets for which the DDS&OP team can configure the DDOM capability. The DDS&OP process then returns model projection and strategic recommendations as part of the Adaptive S&OP scenario evaluation process. Within the DDS&OP process, tactical exploitation influences the tactical demand plan.

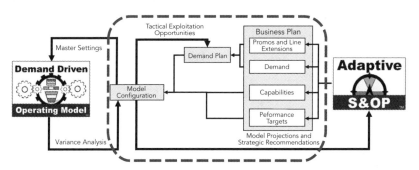

FIGURE 5-1   DDS&OP schema

The Elements of Demand Driven Sales & Operations Planning

There are six basic elements to effective DDS&OP. In order to demonstrate these six elements we will use a sample environment. The environment is not very complex, but it will allow the demonstration of DDS&OP within the short amount of space allotted in this text.

Figure 5-2 depicts a company called "SampleCo." SampleCo is a mixed-mode operation with two make to stock products (FP1 and FP2) and a custom line of make to order products called FPC. Each FPC is somewhat unique but is made through a common set of resources, some of which are shared with FP1 and FP2.

Displayed on the left side of Figure 5-2 are the routings for each product containing the sequence of operations, the resource assignment of the operation, and the time per unit to accomplish each step of the sequence. SampleCo has five resource centers; each one is available for 960 minutes per day (two eight-hour shifts), five days per week only. The routing also contains the selling price for each finished item as well as the direct material prices contained in each unit.

On the right side of Figure 5-2 is the product structure for each item including lead time category assignments and decoupling point placements. The reader can assume a one-to-one parent to component ratio for all levels. Additionally, critical buffer profile definitions are provided for this environment. When a part is decoupled, the decoupling point buffer (DDMRP buffer) will utilize these inputs for buffer sizing. Both FP1 and FP2 have been given the assignment of "MMM," which means manufactured item, medium lead time, and medium variability.

Finally, Figure 5-2 can also help us define the tactical relevant range. The cumulative lead time for FP1 and FP2 is 35 days. In this environment that translates to seven working weeks. For FP2 we see a slightly longer cumulative lead time of 38 days. For simplicity we will use 35 days in the past and future as the tactical relevant range for all products and resources.

SampleCo data generated over the past tactical relevant range will be used to demonstrate some aspects of DDS&OP. Additionally, SampleCo

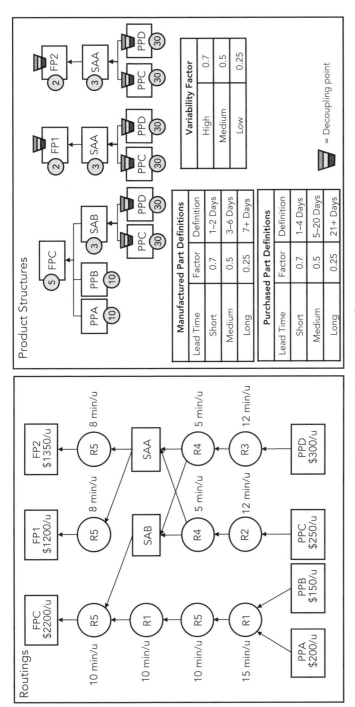

FIGURE 5-2 SampleCo information

scenarios about the future will demonstrate additional aspects of DDS&OP. First, Tactical Review.

Tactical Review

The first element of DDS&OP looks at the past performance of the Demand Driven Operating Model with regard to its reliability, stability, and velocity. These variance analyses identify areas or processes in the DDOM that jeopardize performance, cause additional spend, or present improvement opportunities for model refinement. Typically, these analyses are done through Pareto, run, and control charts.

Pareto analysis charts identify the frequency and severity of outlier observations in buffer performance over time. Figure 5-3 depicts a conceptual Pareto analysis for each type of buffer. In each case, the DDS&OP team is looking at the outer tails of the distribution on either side. The disperson of the tails will indicate how stable the buffers have been. Additionally, when combined with reason code assignments, the DDS&OP team can identify specific action plans for overall performance improvement.

Figure 5-4 illustrates a run chart for the on-hand performance of a stock buffer over the last 30 days. The performance appears to be excellent with regard to the anticipated on-hand range. Rarely does the buffer go above or below the expected range nor is there any evident trend. Furthermore, on the rare occasions it does go outside of the expected range it is minor and recovers quickly. This is a buffer that can most likely be safely adjusted down since the current level of safety appears to be too high. This adjustment would directly lead to a lower amount of inventory investment dedicated to this buffer.

Now we will move from the conceptual view to the SampleCo data generated over the past tactical relevant range. These types of views are designed to be highly visual, intuitive, and impactful for adaptive agents of a complex adaptive system.

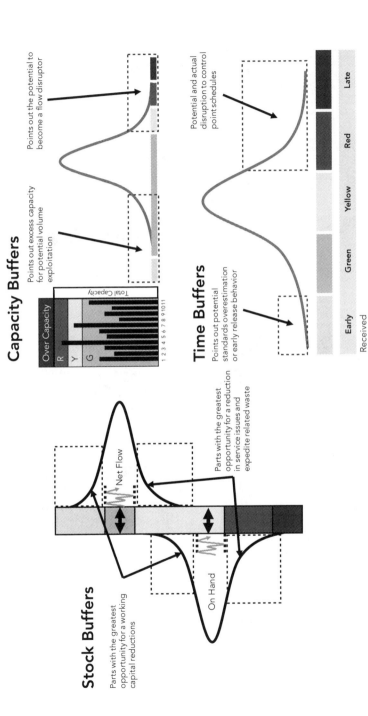

FIGURE 5-3   Pareto analysis of stock, time, and capacity buffers

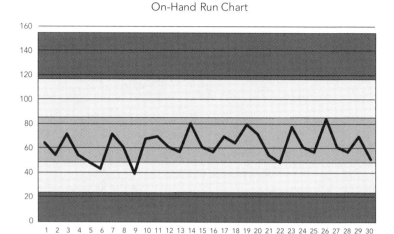

FIGURE 5-4   Stock buffer run chart (on-hand performance)

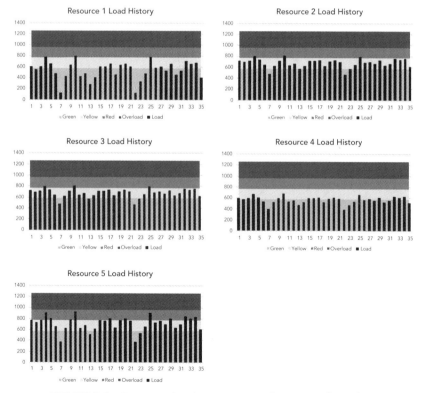

FIGURE 5-5   Resource load patterns over last 35 working days

Figure 5-5 shows resource loading patterns over the past tactical relevant range. Each bar indicates the load of actual demand placed on each resource for each day in the past. The display also indicates how that load compared against the capacity for each resource. The bottom green band represents a 60 percent load for that day's capacity. The next band (yellow) represents 61–80 percent of the resource capacity. The third band (red) is 80–100 percent of daily capacity. The top dark red band is the overload area. Resource load graphs were discussed in Chapter 4.

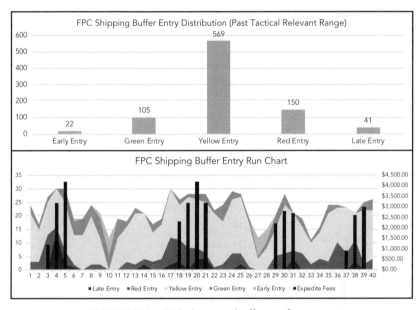

FIGURE 5-6   FPC shipping buffer performance

Figure 5-6 displays the time buffer entries for 887 FPC units shipped over the last 40 days. Forty days was used in order to cover the full cumulative lead time of FPC. The top graphic is an aggregate distribution of buffer entries over this time period. The lower graphic displays buffer entries by day as well as daily additional expedite expense. The graphic is stratified into five colors; the lowest is a dark red color. The lower dark red "spikes" represent late entry to the buffer and consequently a late shipment. The next band is a lighter shade of red and represents a red zone entry. The next bands, in order, depict yellow, green, and early entries. The black bars correspond to the right-hand axis and depict incremental daily spend for expedite including overtime, customer penalty, and/or expedited outbound freight. Time buffers and buffer entry were discussed in Chapter 4.

Figure 5-7 is a run chart showing on-hand position by day versus the on-hand target range for each of the finished product buffers. The on-hand target range is the green band in the middle of the charted data. Above and below that green band are yellow bands that represent a warning zone; either on-hand is becoming too low or too high. Above the upper yellow band is a red band. When on-hand is here, inventory is too high. Obviously, a prolonged on-hand position in this band means a persistently excessive position. Conversely, there is a red band below the lower yellow band. On-hand positions in the lower red band will trigger an "On-Hand

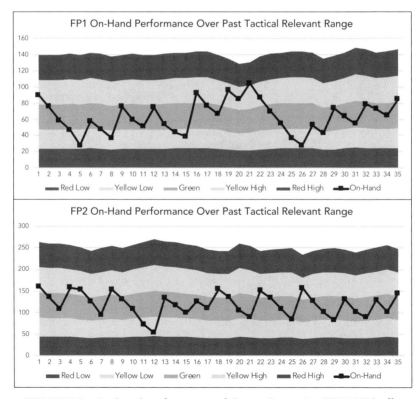

FIGURE 5-7   On-hand performance of decoupling point (DDMRP) buffers

Alert" in DDMRP systems, meaning inventory is too low; the integrity of the decoupling point is being threatened. An on-hand position below the red band means a stock-out with demand; flow has been disrupted. Similar charts would be developed for all decoupling point buffers.

The target range is established through a relatively simple formula. The lower limit of the range is defined by the top of the red zone value. The upper limit of the range is the sum of the red and green zones. Typically, the lower yellow band is one-half of the planning total red zone but there are exceptions and options. The upper yellow band is typically defined by the value of the upper target range minus the top of red plus green zone value. Readers wanting to know more about the specifics of these values should read Chapter 12 of *Demand Driven Material Requirements Planning* Version 2 (Ptak and Smith, Industrial Press, 2018).

With all these data, SampleCo's DDS&OP review will focus on three situations that warrant review.

The Resource 5 Review

Figure 5-5 shows one resource that might be cause for concern from a capacity perspective. Actual demand is pushing its load frequently into the red capacity buffer zone. Why would this be of concern? Remember, Resource 5 is the last operation in the routings. For the FPC product line, Resource 5 directly feeds into the shipping time buffer; it is the last operation before shipment.

If Resource 5 is behind schedule, then the shipping buffer will be impacted. Typically, resources slip behind for one or a combination of three reasons:

• Work orders arrived late to that resource
• That resource is capacity constrained
• Maintenance or break down issues

When all three of these factors are present, expect resources to fall dramatically behind. For simplicity, we have not factored in maintenance or machine reliability in the SampleCo data; the reader should assume the resource centers are always available to run within their shift.

Resource 5, however, does have two of the three factors. Since it is the last resource in the routings, variability will accumulate to often deliver work behind the scheduled time. There is a basic rule in most manufacturing and project environments: delays accumulate, gains do not. The more dependencies involved, the greater the delay accumulation. In fact, this basic rule is the justification for using time buffers.

Resource 5, as evidenced by the loading graph in Figure 5-4, is known as a "capacity constrained resource." A capacity constrained resource simply means that it has very little excess capacity; any mismanagement could result in it becoming a bottleneck. A bottleneck by definition is any resource where capacity is less than demand.

The FPC Shipping Buffer Review

The second situation that warrants review is one that is connected to the first identified issue. When we study Figure 5-5 we see the impact of Resource 5's position in the routings as well as its capacity constrained nature. While the distribution chart looks relatively normal, we must be careful not to overlook specific periods of time that cause disruptions to flow and additional spend.

The lower run chart in Figure 5-5 shows that there are times when the shipping buffer is acutely threatened and/or impacted. As a direct result, we can see a corresponding spend in overtime and fast outbound freight to attempt to close that gap. Over the eight week period a total of $35,700 has been spent in overtime ($17,800) and fast outbound freight ($17,900).

The FP2 Buffer Review

Now we will examine the third situation that warrants review. In this case it is the absence of variance that justifies a review. Figure 5-6 shows that the FP2 on-hand position over the last seven weeks has been consistently in or very near the target range. On only four occasions did on-hand even move into the low yellow band; two of those occasions were extremely slight duration incursions that immediately self-corrected. This kind of analysis can be performed on a large part database by calculation of the statistical capability index also known as Cp and

Cpk. The use of these well-defined statistical tools provides an excellent method to identify potential buffer resizing based on the demonstrated variance of the process over time.

Tactical Projection

DDS&OP is not just about reviewing the past. The DDS&OP process projects model performance given different scenarios within the future tactical relevant range (one cumulative lead time into the future). In order to perform these projections, the DDS&OP team must have an awareness of current and potential problems regarding capacity, supply disruptions, quality/yield issues, and anomalous sales activity.

The DDOM Master Settings can and should be adjusted based on known or planned events taking place within the future tactical relevant range. Examples might include

- Seasonality
- Promotions
- A planned shut-down
- A known supply disruption
- The cut-over to a new manufacturing process
- The impending opening of a new distribution facility
- The introduction of a new product

In the case of SampleCo, there is currently a plan to shut down the shop area where Resources 2, 3, and 4 are located for electrical upgrades four weeks from now. The shutdown will last approximately three days. Three days without those resources functioning will cause a collapse of all stock buffers at higher levels.

With the Tactical Review and Projection complete, the DDS&OP team is now ready to make changes to SampleCo's DDOM.

Tactical Configuration and Reconciliation

The next element of DDS&OP is called Tactical Configuration and Reconciliation. This element is comprised of actions derived from the

Tactical Review or based upon knowledge of upcoming plans within the business. DDS&OP configures the DDOM to match the evolving business plan range and environment through the DDOM Master Settings. Remember, the DDAE model does not utilize a Master Production Schedule. Instead it maintains the capabilities of the DDOM through Master Settings. Figure 5-8 lists the DDOM Master Settings managed by the DDS&OP process. Changes to these settings allow the DDOM to adapt operational capability within a defined range to changing circumstances of the business.

DDMRP	Demand Driven Scheduling
Stock Buffer Profiles. The groupings and settings for replenished parts (part type, variability and lead time).	**Time Buffer Profiles.** The time buffer groupings and settings for scheduled parts.
Planned Adjustment Factors. The adjustment factors to be applied to buffered items or groups of buffered items.	**Time Buffer Profile Assignment.** The assignment of scheduled parts to time buffer groupings.
Part ADU. The average rate of use for each replenished parts.	**Resource Assignment.** The assignment of a resource to a resource type (control point, resource, buffered resource).
Part Profile Assignment. The assignment of each replenished part to particular buffer profile.	**Resource Properties.** Applicable scheduling properties for each resource (capacity, calendar, shifts, operators, etc.).
	Part Properties. Applicable scheduling properties for each part (routings and run rates).

FIGURE 5-8   DDOM master settings

After the Tactical Review of SampleCo data, what actions might we expect the SampleCo DDS&OP team to take with regard to the DDOM Master Settings in Tactial Configuration and Reconciliation?

Protecting Resource 5 and the FPC Shipping Buffer

The Tactical Review of Resource 5 and the FPC Shipping Buffer clearly showed an issue requiring attention. Despite shipping the FPC line at 95 percent on-time, SampleCo is expending significant additional resources (overtime and fast outbound freight) in order to keep up that service level. The DDS&OP team is concerned about this persistent situation especially if FPC sales rise. Something that should be a good thing could turn into a customer service disaster and/or cause a balloon in non-budgeted expedite-related expenses.

The DDS&OP team identified the following options:

Option #1. Declare Resource 5 a control point. Making Resource 5 a control point would mean that it would be finitely scheduled to meet market demand. Then all other subsequent resource schedules would be synchronized to that schedule. Additionally, this would mean the installation of an appropriately sized time buffer in front of Resource 5. Installing a time buffer in front of Resource 5 better protects its schedule since variability passed on by upstream resources can be mitigated.

Option #2. Add decoupling points at PPA, PPB, and SAB. By decoupling those components, the variability on the supply side can be mitigated. For PPA and PPB that means that any supplier variability will be absorbed by the stock buffer. For subassembly B (SAB) it means that accumulated delays from Resource 2, 3, and 4 will be absorbed. That will allow any time buffers after the decoupling points to be minimized.

The team assembles a snapshot of these three new buffer positions based on their part properties. Figure 5-9 shows these three new buffers. The black line represents the calculated average on-hand inventory targets. For both PPA and PPB the average on-hand commitment will be 223 units for each. This represents $48,950 and $33,375 in working capital, respectively. For SAB the average on-hand commitment is calculated as 59 units with a direct material value of $32,450. All buffers together represent an average working capital commitment of $114,775.

It is important to understand that under the current operating model there is still inventory in the FPC line—it is simply coded as work in process instead of decoupled inventory. There are currently 322 open orders on the FPC line. Of those orders, 160 have already committed all components necessary for FPC, with $144,000 in WIP. The remaining 162 orders have already committed components PPA and PPB ($56,700 in direct material value). The total work in process inventory on this particular day is $200,700 in committed direct material.

A proper comparison between the current situation and the proposed situation has to compare the current working capital position with the proposed working capital position, including work in process. Under the proposed buffering change there would be an additional $70,800 in work in process inventory in FPC assembly (using the last five days of sales) as well as two replenishment orders in process PPA and PPB and one replenishment order in process for SAB. This adds an additional $48,840, $33,300 and $35,200, respectively. This brings the total stocked and work in process inventory commitment under the proposed scenario to $232,115, representing an increase of average committed working capital of $31,415.

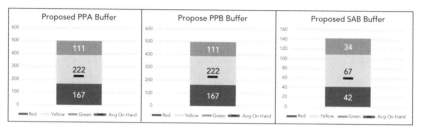

FIGURE 5-9   PPA, PPB, and SAB proposed buffers

Decoupling these additional points also means a significant drop in lead time for FPC from the current 15 days to a potential 5 days. The DDS&OP team considers what that might mean for sales opportunities. If the market currently accepts a 15-day lead time, what would be the advantage of a five-day lead time? The answer to that question will be covered later in this chapter.

Option #3. Increase the size of the FPC shipping time buffer. With additional time in that buffer, it would allow more variability to be absorbed and still provide on time delivery without fast freight or overtime.

After some deliberation, the DDS&OP team decides to immediately adopt consideration 1 and ask for immediate approval from management for adopting consideration 2. The third consideration was deemed unnecessary given the adoption of the previous two. Figure 5-10 shows the updated SampleCo model. Decoupling points have now been placed at PPA, PPB, and SAB. Resource 5 is highlighted to show that it is now a control point.

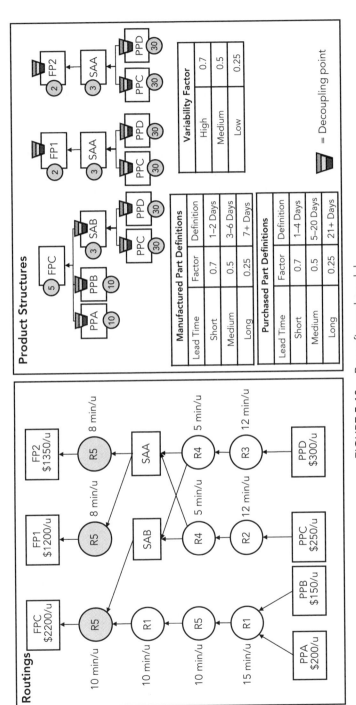

FIGURE 5-10 Reconfigured model

FP2 Stock Buffer Migration

As mentioned in the Tactical Review section, the FP2 stock buffer is performing so well that it presents an opportunity to reduce working capital. The DDS&OP team decides to move the FP2 stock buffer from the medium variability profile to the low variability buffer profile. This allows inventory dedicated to this buffer to be reduced.

Figure 5-11 shows the difference to the FP2 buffer resulting from the buffer migration. The only buffer zone affected is the red zone, which moves from a value of 83 to 69. The entire red zone value is part of the average inventory equation. This change to the buffer results in a reduction of $7,700 in average on-hand inventory (14 × $550).

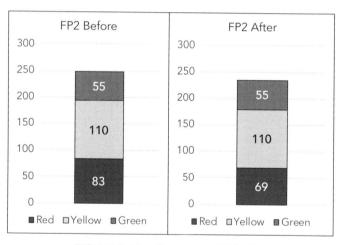

FIGURE 5-11   Change to FP2 buffer

Now the DDS&OP team will simulate the last seven weeks with this new buffer profile for FP2. Simulation is a vital aspect of successful DDS&OP; it can better enable both the "sensing" and "adaptation" aspects of a model. Figure 5-12 is the simulation results for the new FP2 buffer configuration assuming the last seven weeks of demand. In this particular case, the simulation looks advantageous to make this configuration change.

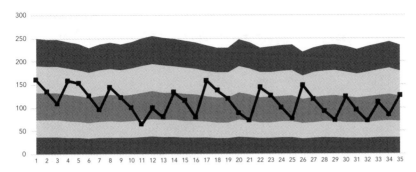

FIGURE 5-12   Simulated FP2 on-hand performance

Dealing with the Future Shutdown

As discussed in the Tactical Projection section, there is a plan to shut down Resources 2, 3, and 4 for three days scheduled in four weeks. The DDS&OP team formulates a plan to attempt to minimize the impact of this shutdown. They will temporarily inflate the FP1, FP2, and new SAB stock buffers through a type of planned adjustment known as a "zone adjustment factor."

Inflating the SAB buffer will allow the FPC product line to continue production through the shutdown. Resources 1 and 5 are unaffected by the shutdown. Inflating the FP1 and FP2 buffers will allow finished product sales to continue through the shutdown.

The planned shutdown will last three days. The DDS&OP team plans to inflate each buffer three times the average daily usage of each buffer. Today's average daily usage (ADU) for FPC is 22.2. That means that the SAB buffer must be inflated by 67 (22.2 × 3) additional on-hand units. That will be relatively easy given the spare capacities of Resources 2, 3, and 4.

FP1 and FP2 will be more difficult because of the constrained capacity of Resource 5. Today's ADU for FP1 and FP2 is 13 and 22, respectively. That will require an additional 39 FP1 units and 66 FP2 units. This additional work will require a total of 840 minutes of capacity for Resource 5, nearly two shifts of capacity. The plan will be to work a mandatory full shift of overtime for both shifts on the Saturday before the shutdown.

This is just one simple example of a DDS&OP process configuring the model not based on what has happened but on what is planned or expected to happen. Now consider the next element of DDS&OP called Tactical Exploitation.

Tactical Exploitation

DDS&OP can bring short range supplements to flow when/if capacity is available as well as look for ways to minimize cash outlays while maintaining and/or increasing flow. Tactical exploitation requires visibility to DDOM capability in the short term in order to be opportunistic in nature within the relevant time frame. The DDS&OP team has the proverbial keys to the car for the short range; they drive it and maintain it. Given the provided capacity, how can they best use its capability within the tactical range?

One SampleCo example of Tactical Exploitation has already been covered. The declaration of Resource 5 as a control point and the installation of decoupling point buffers at SAB and PPA and PPB were an attempt to stabilize the FPC shipping buffer and to reduce the amount the company is spending on overtime and premium outbound freight. Yes, establishing the decoupling point buffers takes the commitment of some additional working capital, but it was noted that there was still a significant amount of inventory already in the FPC line as work in process. Additionally, if those stock buffers deployed in concert with the Resource 5 control point can significantly reduce these expenses, that recovered cash allows the FPC line to accomplish the same with fewer resources and less spending.

But what about doing *more* with the same or fewer resources? Is there additional capacity available that can be sold without affecting the flow of existing business through the environment? For example, if a company's heat treat facility is scheduled well below capacity for the week, are there local companies that would pay for that heat treat capacity? Anything charged above truly variable costs results in additional cash with the same fixed cost base. Remember, labor cost is most likely part of that fixed cost base within the tactical relevant range.

In SampleCo there are some potentially promising directions in this regard. First, perhaps SampleCo can find a company that needs the capability of Resource 1 and/or 4? Both have spare capacity and could take on additional business if an outlet for that capacity can be found.

The second opportunity involves the capabilities of the FPC line. As described above, the DDS&OP team has asked for approval to add new decoupling points at SAB, PPA, and PPB and is adding a control point at Resource 5 in order to better protect the FPC shipping buffer performance. The placement of those decoupling points dramatically reduces the decoupled lead time of FPC from 15 days to 5 days. If the market currently accepts 15 days (SampleCo ships 95+% at that lead time), then SampleCo should be asking about segmenting market opportunities to get a higher selling price for shorter lead time.

The DDS&OP team must be careful for two important reasons:

- Telling the Sales team that the new lead time is now five days could result in a flood of volume that would quickly overwhelm Resource 5's total capacity. While SampleCo would immediately take a lot of market share, it would then quickly squander that opportunity from a customer service objective and pay dearly for the overtime and expedited freight.
- Why give something away for free? If SampleCo cannot take all of the potential volume (at least in the short term) by moving to a five-day lead time, then it must selectively apply its agility to its best advantage. That advantage could be used to command a higher price for certain quick turn orders or to perhaps win or retain a critical account.

This example shows the importance of having Sales and Marketing representation on the DDS&OP team. What are the market possibilities with a five-day lead time for the FPC line? As it turns out, there is a small market for quick turn opportunities, particularly due to last-minute changes in the projects in which the FPC product is typically used. As evidence, Sales cites one particular customer who is frequently asking for shorter lead times with these instances. They have even paid

SampleCo's overtime in the past to work a Saturday shift and paid a 15 percent expedite upcharge.

The DDS&OP team decides to allocate 15 units of FPC production to an expedite lane. It will be up to Sales to best use that allocation for the benefit of the company. That could mean premium prices (adding more contribution margin) and/or winning a new account (adding additional volume). It is also emphasized that all 15 of these slots should not be sold if there is no incremental advantage to SampleCo (the customer would have accepted the 15 day lead time).

Additionally, the DDS&OP team explores the possibility of producing more units in five days while using less expensive but slower freight options to certain regions of the country. The customer still experiences the 15-day lead time but SampleCo saves on total freight expenditure.

It should be noted that sometimes, in order to generate additional cash, it will require an investment of cash. Thus, the DDS&OP team is responsible for variable costs (overtime, freight, material, etc.) associated with operating the DDOM in the tactical relevant range. They must best judge how to use and complement the fixed asset base given the existing sales and potential opportunities to which they have visibility.

Strategic Recommendation

The DDS&OP team presents ideas for better DDOM performance needing senior-level approval to the management team. These could be internal innovations designed to make the DDOM faster, more stable, or more reliable, but their approvals are above and beyond the authority of the DDS&OP team. An example might be the recommendation for a third shift, a new piece of equipment, or the reengineering of a specific product to be manufactured differently. While these recommendations are beyond the DDS&OP team authority, the DDS&OP team typically has the best ability to make a justifiable case for change and relate the options to both tactical and strategic metric objectives.

This is why it is imperative that the DDS&OP team has an excellent working knowledge of the resources and products that flow through the model. Both of these can be manipulated in the long term to increase total flow through the model.

Turning our attention back to SampleCo, the DDS&OP team must ask itself what the biggest obstacle to future flow might be. While they have taken some short-term actions to help Resource 5 deliver consistent performance, growth for any of the products (FPC, FP1, or FP2) will create further strain on that resource. How can SampleCo plan to address this issue in the longer term?

There are at least two intriguing options:

Option 1. Add more resources by recommending a third shift for Resource 5. A third shift would add 480 minutes of Resource 5 availability per day. It would also add to the fixed expense base; operators, support staff, and maintenance would all need to be present. This would present some immediate challenges:

- Finding and retaining qualified operators for a night shift may prove to be extremely challenging.
- Adding 480 minutes of Resource 5 per day to enable increased production is not the whole story. Would that addition really result in 50 percent more total volume potential for the entire plant, or would another resource simply become a bottleneck? The most likely candidates would be Resources 2 and 3. When the DDS&OP team does the analysis, Resources 2 and 3 definitely are next in line to become the next capacity constrained resources. Figure 5-13 illustrates this analysis. According to Sales, the biggest current growth potential is FPC (especially with better lead time). The DDS&OP team first decides to simulate a scenario with a 50 percent increase in FPC business while other business lines stayed constant. Figure 5-11 shows that the plant should be able to absorb that volume, but Resources 2 and 3 will be stretched thin from a capacity perspective. The team then decides to simulate a modest 15 percent growth for the FP1 and FP2 lines. This analysis shows that Resources 2 and 3 will be further stretched. Because the demand numbers feeding the analysis are averages, any per-

centage load over 90 percent almost guarantees that there will be occasional, if not frequent, bottlenecks for these resources as daily demand (especially on a make to order product) will fluctuate widely. Still, this seems to be a viable long-term option assuming the volume growth potential is there.

	R1	R2	R3	R4	R5
Current load (minutes)	554.4	686.1	686.1	571.8	723.5
Current load %	57.7	71.5	71.5	59.6	75.4
FPC up 50%	831.6	819.2	819.2	682.6	945.3
FPC 50% up load %	86.6	85.3	85.3	71.1	65.6
FPC 50% up & FP1 & FP2 up 15% load (minutes)	831.6	882.2	882.2	735.1	987.3
FPC 50% up & FP1 & FP2 up 15% load %	86.6	91.9	91.9	76.6	68.6

FIGURE 5-13   Resource 5 third shift analysis

Option 2. A redesign of SAA to reduce R5 assembly time. The original designs of FP1 and FP2 were done with no concern about assembly time. A DDS&OP team member with an in-depth understanding of both products, as well as the manufacturing capabilities to make them, has come up with an idea. An additional step can be added to the operation for SAA that will cut the assembly time on Resource 5 by two minutes per piece for FP1 and FP2. Engineering has verified that the idea could work and the operators have also verified that Resource 4 is process capable to handle that extra step. It will, however, require three minutes of additional processing on the PPD leg of the routing.

Figure 5-14 shows the impact of this proposed design change. While it does add a larger proportion of load onto the R4 resource, it will provide more R5 capacity availability with no additional spend. Combined with R5 being a control point with a time buffer and the fact that R5 is the last in the sequence, this idea seems to have some merit for the next six to twelve months. Engineering expects this change to take roughly six to eight weeks to implement. The engineering cost is not relevant because the engineering department is a fixed cost.

The DDS&OP team decides to pass both options to the Adaptive S&OP team for strategic review. Shortly after delivering these recommendations they get a request from the Adaptive S&OP team to evaluate the impact of a major proposal being worked on at the highest levels of the company. This example brings us to the final element of DDS&OP: Strategic Projection.

	R1	R2	R3	R4	R5
Current load (minutes)	554.4	686.1	686.1	571.8	723.5
Current load %	57.7	71.5	71.5	59.6	75.4
SAA change load (minutes)	554.4	686.1	686.1	676.8	688.5
SAA change load %	57.7	71.5	71.5	70.5	71.7

FIGURE 5-14   SAA redesign impact

Strategic Projection

The DDS&OP process also helps project potential outcomes in the strategic relevant range based on the scenarios being evaluated in the Adaptive S&OP integrated reconciliation process. DDS&OP provides a feasibility check from an operational perspective in the Adaptive S&OP process (which will be discussed in Chapter 6). In short, DDS&OP's job in the Adaptive S&OP process is to bring operational reality and/or possibility to the strategic plan through Strategic Projection. This is a vital aspect of the Adaptive S&OP integrated reconciliation process. Our SampleCo example will continue with a demonstration of this output and the corresponding projection.

A foreign entity has approached SampleCo and proposed a joint venture. The proposal would have SampleCo producing and shipping SAAs to a foreign site for final assembly of FP1 and FP2. By assembling in the foreign nation, the joint venture will avoid certain prohibitive trade restrictions as well as qualify for investment matching funds from the foreign government. This is a market that has previously been closed to SampleCo and could prove to be a template for additional joint ventures in other foreign countries.

The DDS&OP team's job is not to debate whether there are intellectual property issues or whether this fits into the company's overall strategy; that is the role of the Adaptive S&OP process. The DDS&OP team's job is simply to evaluate the impact this proposal will have on the DDOM and/or how the DDOM must change to accommodate this strategy with the relevant financial impact.

The Adaptive S&OP team communicates this scenario:

- The foreign entity believes the potential market to be immediately significant. Initial projections are an average monthly sales rate of 100 for FP1 and 160 for FP2. For reference, domestic sales of FP1 are currently averaging 260 per month and 440 per month for FP2.
- The selling price to the foreign entity would be $750 per unit for both FP1 and FP2.
- The proposal is to begin production at the foreign site within eight months.
- It will be critical to control shipping costs in order to maintain overall profitability.

Now the DDS&OP team needs to go to work. They have some immediate impressions based on the current configuration of the DDOM:

Decoupling Points

This new scenario will certainly require at least one additional decoupling point, perhaps two. First, the team explores the potential for a SAA decoupling point at the domestic factory. The domestic demand is combined with the potential foreign demand to produce a projected buffer size at SAA. The placement of the decoupling point at SAA also has an impact to the FP1 and FP2 buffers because it reduces their decoupled lead time. Figure 5-15 shows the impact of buffering SAA at the SampleCo factory, including the average working capital commitment. The row labeled ADU is determined by dividing the sum of weekly sales of both FP1 and FP2 by five. The column labeled "SAA

Domestic Buffer" has an ADU that includes the foreign sales of FP1 and FP2 since it will be the source of material for those units. This SAA domestic buffer will require an average working captial commitment increase of $111,650.

SAA Domestic Buffer		SAA Foreign Buffer	
ADU	48	ADU	13
Lead Time Factor	0.7	Lead Time Factor	0.25
Variability Factor	0.5	Variability Factor	0.5
Domestic SAA Lead Time	3	Foreign SAA Lead Time	30
MOQ	0	MOQ	100
Red	152	Red	147
Yellow	144	Yellow	390
Green	101	Green	100
TOG	397	TOG	637
Average On-Hand	203	Average On-Hand	197
Direct Material	$550	Direct Material	$550
Average On-Hand $	$111,650	Average On-Hand $	$108,350
Total Increase			$220,000

FIGURE 5-15   SAA buffer impact

Figure 5-15 also includes the consideration of another decoupling point at the foreign factory (labeled "SAA Foreign Buffer"). This would allow for SAAs to be available on demand to production. This would remove the necessity to tie SAA ordering to forecast at the foreign factory; this is something the DDS&OP team would like to avoid. This column shows an additional $108,350 in average working capital required. Transportation lead time is set at 30 days. Additionally, there is a minimum order quantity set at 100 (enough to fill a 20-foot container). While it appears that the transportation lead time requires a fairly significant inventory commitment, one has to realize that inventory is going into the system regardless of whether it is in a decoupling point buffer or strung out in multiple shipments tied to forecast.

Combined with the domestic SAA buffer, there will be a total average increase of $220,000 in working capital dedicated to both of these positions. The decoupling point buffer at the domestic SAA position, however, does allow for a reduction in the FP1 and FP2 decoupled lead times. Both positions' lead times drop from five days to two days. Figure 5-16 shows the reduction in the FP1 and FP2 stock buffers. Figures 5-15 and 5-16 show a net increase in average working capital for the company of $192,768. It is worth noting that FP2 includes the reduced variabilty factor mentioned earlier in this chapter when discussing Tactical Configuration and Reconciliation. Maintaining this variability factor change is justifed since the establishment of the SAA domestic buffer will further reduce any supply variability experienced by FP2. Furthermore, the DDS&OP team has determined that under this scenario it is justified in moving FP1 to this lower variability profile given the same reasoning.

	FP1 Current	FP1 Projected	FP2 Current	FP2 Projected
ADU	13	13	22	22
Lead Time Factor	0.5	0.7	0.5	0.7
Variability Factor	0.5	0.25	0.25	0.25
Domestic SAA Lead Time	5	2	5	2
Red	49	38	69	53
Yellow	65	36	110	60
Green	33	26	55	21
TOG	147	100	234	134
Average On-Hand	49	38	97	64
Direct Material	$550	$550	$550	$550
Average On-Hand $	$26,983	$17,626	$53,075	$35,200
Reduction		$9,357		$17,875

FIGURE 5-16 Impact to FP1 and FP2 positions

Resources

Both the establishment and maintenance of the buffers will impact some resources dramatically. While this proposal does not require any domestic assembly capacity (Resource 5), it will push Resources 2 and 3 into being capacity constrained resources. Figure 5-16 shows the load for the involved resources (2, 3, and 4) combining current and proposed average daily usage of SAA as well as current SAB average demand. Figure 5-17 uses the original time per unit for Resource 4 rather than the proposed additional time (the SAA engineering change) from the previous section.

	R2	R3	R4
Current load	686.1	686.1	571.75
Current load %	71.5	71.5	59.6
Projected load	842.1	842.1	701.75
Projected load %	87.7	87.7	73.1

FIGURE 5-17   Resource 2, 3, and 4 loading under proposed joint venture

Figure 5-18 shows the updated DDOM design. Under the proposal, Resource 5 continues to be a control point. The team decided not to make Resources 2 and 3 control points for the following reasons:

- The resources are gating operations being fed directly by a decoupling point buffer. This means these resources are not subject to previous resource schedule variability.
- The Resource 5 control point schedule will have no direct scheduling effect on Resources 2 and 3 schedules because of the decoupling points at SAA and SAB.

Clearly, Resources 2 and 3 can become a concern if actual demand experiences growth beyond the current projection. The DDS&OP team determines this is stretching the model to the limit but should be manageable for the time being with supplementary overtime. Beyond that, the DDS&OP team will work on a proposal for additional capacity. This will most likely involve the use of a limited third shift.

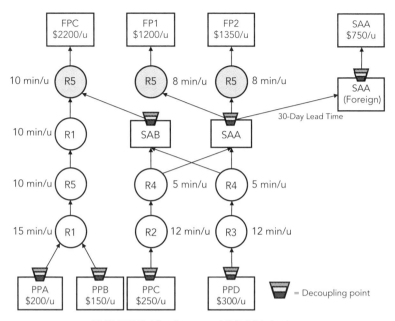

FIGURE 5-18 Proposed DDOM design

One critical consideration, however, is how to establish these buffersed positions within the required time frame. The design calls for an average of 400 SAA units to be held on-hand between the two sites. These buffer positions must be established while normal operations continue. The DDS&OP team proposes to use a mandatory Saturday shift (both shifts) every other week for Resources 2, 3, and 4. Each Saturday these two additional shifts should be able to build approximately 80 SAAs. The SAA units dedicated to the foreign stock buffer position will be built and shipped first due to the long transportation time. The DDS&OP team has calculated that the foreign stock position can be in place within four months. The domestic position target will take an additional month and a half to build.

Financial Implications

This proposal has some significant financial implications relative to the nature of the DDOM. It is the DDS&OP team's job to give those details to the Adaptive S&OP team. We have already detailed the working capital implications of this proposal. Now contribution margin and additional expenses regarding this proposal are considered.

Consider contribution margin first. The foreign entity has agreed to pay $750 per SAA. Each SAA has a direct material cost of $550. This seems straightforward enough—increased cash per unit of $200. But is it the complete story? A variable and relevant expense is the cost associated with shipping. Under the proposal, SampleCo would be responsible for shipping costs. A 20-foot container will average approximately $4,000 per shipment from SampleCo to the foreign factory. A 20-foot container can fit 100 SAAs; that is why 100 is being used as the MOQ. If each container has 100 units and SampleCo gets $200 contribution per unit, then each container contains $16,000 in direct contribution after accounting for shipping costs. This essentially means that each unit going overseas is really only returning $160 cash per unit on average.

As noted above, Resources 2 and 3 will become capacity constrained under this proposal. Additional business growth could cause them to become the production limitation or primary determinant of system output. Any time a system has a limitation, it is critical to understand how that limitation is utilized to maximize the company's profitability. When there are options in how to use a system limitation, those options differ particularly with regard to a rate of return.

One basic tenet of management accounting is that a company will maximize profit when it makes and sells the products or services with the highest contribution margin per unit of the scarce resource. If the limitation ultimately determines what can come out of the system, then a company should want to maximize the rate of return at that limitation in order to maximize the rate of return for the system. One way to judge this is to look at the contribution margin per the amount of usage of the limitation. This gives us a contribution margin per minute on the limiting factor's capacity.

Figure 5-19 is an analysis on Resources 2 and 3 (which have identical capacity properties relative to the products). Ultimately, Resources 2 and 3 only directly contribute to two things, SAA and SAB, but we should think of SAA and SAB as "storage tanks" of capacity. In order to have an

SAA or SAB, SampleCo had to have invested 12 minutes of Resource 2 and 12 minutes of Resource 3. SAB can only be directed to one end item: FPC. SAA can be directed to three distinct market interfaces (FP1, FP2, and the foreign joint venture); all have a different rate of return.

	SAB	FP1	FP2	Foreign
Selling price	$2,200.00	$1,200.00	$1,350.00	$750.00
Direct material cost/unit	$550.00	$550.00	$550.00	$550.00
Additional variable cost/unit				$40.00
Contribution margin/unit	$1,650.00	$650.00	$800.00	$160.00
Load/minute	12	12	12	12
Contribution margin/minute	$137.50	$54.17	$66.67	$13.33

FIGURE 5-19   Contribution margin per minute of Resources 2 and 3

Now it is up to the Adaptive S&OP team to make a decision or ask for further clarity. Is this proposed direction in the long-term interest of the company or will it commit the company to a long-term albatross that will stifle or constrain more lucrative domestic potential? That is a question that only management can answer when considering the strategic relevant range. This will be considered in the next chapter on Adaptive S&OP.

Tactical Metrics

DDS&OP is responsible for measuring the past performance of the DDOM based on the operational metrics of reliability, stability, and velocity. But how should we measure performance in the tactical relevant range?

Tactical metrics emphasize DDOM improvement, waste reduction, operating expense control, and strategic contribution in order to determine relevant information and materials in the tactical relevant range for improved ROI. Figure 5-20 displays the tactical metric objectives and messages.

Metric Objectives	The Message Behind the Objective
Tactical Improvement and Waste Reduction (Opportunity $)	Identify and prioritize obstacles and/or conflicts to flow
Tactical Expense Control	Spend minimization to meet the requirements of the market and the DDOM design
Tactical Contribution	Maximize system return according to relevant model factors and tactical opportunities (volume and rate)

FIGURE 5-20   Tactical metrics

Tactical Improvement and Waste Reduction measures the frequency and size of obstructions to flow as well as the cost to mitigate these obstructions. Is the DDOM performing well and improving with regard to reliability, stability, and velocity over time?

Tactical Expense Control seeks to minimize the total amount of spend within the tactical relevant range in order to meet expectations. In some cases, this could mean additional spending is required to capture desirable market opportunity. Charges for expedites, partial ships, and third-party services should be carefully tracked and made visible over time as these are often directly related to DDOM performance issues.

Tactical Contribution directs the organization to identify and exploit opportunities to improve total flow from both a rate and volume perspective. Are there temporary opportunities available to generate additional cash with current resources? Are there opportunities to improve the rate of cash generation at critical or constrained operations?

Summary

DDS&OP provides that crucial bidirectional link between strategic direction and operational capability through six key elements detailed in this chapter. DDS&OP drives the tactical adaptation cycle in order to create a system that is constantly adapting to what has happened, what is happening, and what is planned or proposed to happen.

DDS&OP mechanisms and metrics are based on physics, management accounting, and complex systems principles that are currently unconventional but at the same time intuitive and appeal to common sense. The key is for the DDS&OP team to have the ability to relay that common sense across the DDAE model and the organization.

But an explanation of DDS&OP cannot be complete without explaining the final component of the DDAE Model—Adaptive Sales and Operations Planning.

CHAPTER 6

Adaptive Sales and Operations Planning

We now turn our attention to the strategic component of the Demand Driven Adaptive Enterprise Model. As previously discussed, companies are facing an ever increasingly complex environment coupled with increasing marketplace volatility. This requires companies to strive to be more flexible and responsive by focusing on flow, streamline both planning and execution, and attempt to conserve the use of their valuable resources. Instead of investing capacity, cash, and space into items prematurely, a company must be able to use its resources to produce only those products that will be actually demanded by their customers. How can a company transform to an agile demand driven enterprise capable of staying ahead of today's hypercompetitive market? Clearly, the answer must be a systemic change—an alignment of strategy and demand driven operations.

What Is Sales & Operations Planning?

Sales & Operations Planning (S&OP) was first developed and named by Dick Ling in 1985 while he was working at Ollie Wight. In 1987, he coauthored the first book on S&OP, *Orchestrating Success—Improve Control of the Business with Sales and Operations Planning* (Ling and

Goddard, Wiley, 1987). Today, while S&OP has been around for over 30 years and has enjoyed moderate success with many companies, recent name changes like "Integrated Business Planning" have rekindled more emphasis on the process.

According to APICS, S&OP is:

> A process to develop tactical plans that provide management the ability to strategically direct its business to achieve competitive advantage on a continuous basis by integrated customer-focused marketing plans for new and existing products with the management of the supply chain. The process brings together all the plans for the business (sales, marketing development, manufacturing, sourcing, and financial) into one integrated set of plans. It is performed at least once a month and reviewed by management at an aggregate (product family) level. (APICS Dictionary, 14th edition, p. 154)

As noted in Chapter 3, in a conventional operating schema characterized by the use of a Master Production Schedule at the tactical planning level and material requirements planning at the operational level, this monthly process will send shock waves down through the operational levels of the organization. Most people within operations have very strong opinions about the value of these monthly updates and the level of reality attached to them. This is what was referred to in Chapter 3 as tactical demolition and reconstruction.

The authors want to stress, however, that S&OP is a necessary and important management process for any company and is essential for a company to become a mature and sustainable Demand Driven Adaptive Enterprise. This requires organizing and utilizing people in the right way, with the proper attitude and motivation. This process must foster an environment of trust throughout the company and instill that in its people. Next is the concept of collaboration. This requires people working together with aligned goals (coherence). In a truly collaborative environment, it is not about who is right, it is all about what is right for the company now and into the future. This is enabled by employing flow-based metrics that align the individual, subsystem, and company goals.

As discussed in previous chapters, the Demand Driven Adaptive Enterprise framework is about enabling the true promise of S&OP, providing it a framework with a context rooted in the reality of flow and true bidirectional collaboration.

The Strategic Relevant Range

As noted previously, there are three relevant ranges that both convention and the Demand Driven methodology recognize: operational, tactical, and strategic. In Chapter 4 we defined the operational relevant range (decoupled lead time). In Chapter 5 we defined the tactical relevant range (at least one cumulative lead time in the past and future). This leaves an obvious conclusion about the strategic relevant range: it must begin at one cumulative lead time in the future. But where does it end?

Figure 6-1 depicts all three relevant ranges. Note that in Figure 6-1, the S&OP process does not encompass all strategic planning. The horizon for S&OP must be at least as long as the lead time for capital investment or other strategic initiatives. S&OP overlaps the strategic planning process and enables a company to do a superior job of constructing the annual business plan.

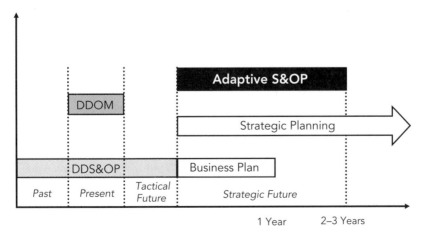

FIGURE 6-1 Relevant ranges in the DDAE Model

As described earlier, Adaptive Sales and Operations Planning (AS&OP) is the integrated business process that provides management the ability to strategically define, direct, and manage relevant information in the strategic relevant range across the enterprise. Market driven innovation is combined with operations strategy, go-to-market strategy, and financial strategy to create strategic information and requirements for tactical reconciliation and strategic projection in order to effectively create the desired future, drive adaptation, and manage change.

Adaptive S&OP is typically performed by the Adaptive S&OP team. The composition and size of this team is relative to the size and complexity of the organization. In major corporations it would be a dedicated team with the following personnel:

- **Adaptive S&OP team leader.** In larger companies the CEO is not the Adaptive S&OP team leader. Instead the team is led by a trusted and respected strategic-minded leader that is still good with detail management and cross-functional communication.
- **Cross-functional C-level direct reports.** Typically, these are the trusted "right-hand" personnel to the executives who are extremely knowledgeable about the reality of the operational environment and their respective functions.
- **DDS&OP team leader.** The DDS&OP team leader provides critical insight to the DDOM capability and the tactical implications for the proposed strategic directions.
- **Adaptive S&OP analysts.** Several may be required to document, integrate, and build communication vehicles around the adaptive S&OP process.

Adaptive S&OP Starting Assumptions

Adaptive S&OP has eight critical starting assumptions. Plainly said, if these assumptions are not in place, Adaptive S&OP cannot and will not operate properly.

- There is a defined strategic direction. Strategy can be expressed or implied, but every company must have a strategy. Senior management must clearly articulate a strategy, or the subsystems will assume what it should be, easily making invalid assumptions (with all the best intentions), thereby potentially pursuing different strategies and risking coherence. The senior management team is responsible for envisioning and articulating the desired future state and how the company will innovate to sustain and grow market pull in the future. The critical questions to ask include
  - How do we view the future economic climate?
  - What relevant emerging trends can affect us?
  - What are our risk mitigation strategies?
  - What should our portfolio look like?
  - What are the economic factors that are affecting us?
  - What will our market share be and what can we do to enhance it?

  Given these factors, the management team must review the current strategy and decide what changes are required, including what projects, customers, territories, and channels should have priority. Strategy, by definition, is not a linear path that goes exactly as expected. Strategy must be reviewed, and adaptation is necessary to achieve the overall top-level business goals. Think of a sailboat tacking to the wind to achieve its final desired destination or in a severe storm adapting to an entirely new destination.
- The Adaptive S&OP process assumes an existing (at least partially established) business plan to begin. The Adaptive S&OP process develops the business plan based on things like:
  - New directives from senior management
  - New opportunities in the marketplace
  - New product timing
  - Deviations from the business plan
  - Supply and DDOM problems
  - Impact of external factors (currency shifts, new regulatory requirements, global crises, competitive threats, etc.)

- The future will look different from the past. The VUCA world we operate in today requires that we understand the nature of complex systems and the dangers associated with making future commitments and decisions based on the past repeating itself. This book has given at least two examples of conventional management practices making this dangerous assumption: the use of fully absorbed unit cost to drive operational metrics and future decisions, and the use of historically derived forecasts to directly launch supply orders.

- The basic difference between managing for flow and managing for cost is understood by the organization. An organization that has not made the conversion to flow-based efficiency (as opposed to cost-based efficiency) simply cannot sustain an Adaptive S&OP process.

- The organization has at least a partial flow-based operating model (DDOM) in place. Without a flow-based operating model, one of the prerequisites for relevant information, operational performance and potential continues to be subject to unrealistic assumptions.

- The organization will have the capability and personnel to perform tactical reconciliation activity to the flow-based operating model (defined DDS&OP activities). Without the ability to tactically reconcile and reconfigure a DDOM, innovation, adaptation, and system efficiency will be stifled in the short and long term.

- Information should be presented and understood as a roughly right range rather than precisely wrong discrete numbers. The DDOM and DDS&OP process lends itself to this basic requirement very well since discrete buffer sizing numbers always translate to ranges of buffer capability and performance.

- You cannot understand Adaptive S&OP without understanding the Demand Driven Adaptive Enterprise model and its characteristics in relation to complex adaptive systems. Chapter 7 provides a review of these characteristics. While many Adaptive S&OP activities (described later in this chapter) seem the same or similar, there are critical differences. It is important to remember that DDS&OP has a critical role in the strategic recommendation process. A company does not implement a Demand Driven Adaptive Enterprise model directly. It must first implement an operating model with DDMRP at its core, then implement DDS&OP to manage the tactical range and then integrate this with the Adaptive S&OP process steps to transform the overall management processes and therefore become "adaptive." This sequential development process is detailed in Chapter 8.

The Seven Steps of Adaptive S&OP

The Adaptive S&OP process manages the strategic relevant range in a series of seven steps as shown in Figure 6-2. This chapter will focus on the seven steps inside the dashed box. This overall process is the realization of the original vision for S&OP as the integrated business process that provides management the ability to strategically direct its businesses to achieve competitive advantage on a continuous basis by the protection and promotion of return on investment. Product innovation, customer focused marketing plans for new and existing products, operations strategy, and the financial strategy are managed on a continuous basis to enable the company to sense, adapt, and innovate successfully across the supply chain.

It is vital to understand that these seven steps and the assumptions involved with them should be carefully documented. Communicating numbers is always necessary but what is more important than numbers is the rationale or logic behind the numbers. If the numbers turn out to be wildly incorrect, typically it means at least one assumption was invalid. Documenting and displaying the assumptions with the numbers builds a critical aspect of trust. The numbers cannot be seen as simply appearing from thin air. If the numbers change, it is important to document why they are changing. Furthermore, by clearly stating the assumptions, an opportunity arises for critical feedback from someone who may have knowledge that invalidates the assumption.

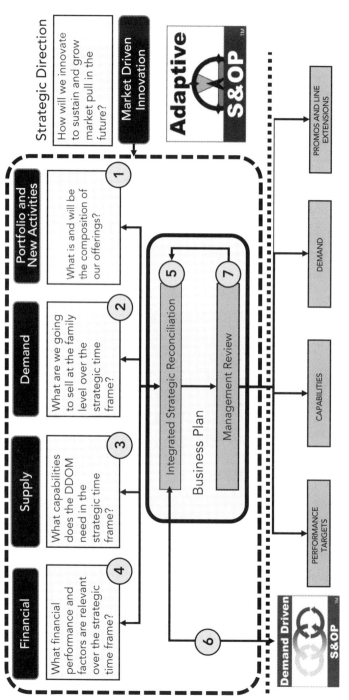

FIGURE 6-2 The Adaptive S&OP schema

Step 1—Portfolio and New Activities

A critical question of Adaptive S&OP is, "What is and will be the composition of our offerings?" Key to any S&OP process is the definition and understanding of product families. Product families can be aggregated at the sales family, marketing family, manufacturing family, or financial family level. The Adaptive Sales and Operations Planning process integrates and reconciles these views with the financial perspective. The use of analytics allows the management team to see different views of the data using the same reference source.

The emphasis during the S&OP process is on product families, not on specific SKU level forecast. One exception may be in the new product area where the focus could indeed be on a specific SKU. Each department usually wants to look at things a bit differently, so there is a need for translation so that the information is meaningful and aligned. Figure 6-3 describes some of the different product family views required and the considerations for each view. Functional area product family definition allows a rough-cut view of this information to provide a high level view of the overall feasibility of the business plan from each perspective. Recent advancements in technology have made this data management easier.

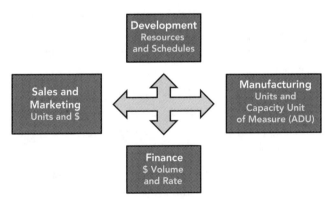

FIGURE 6-3   Translation required for product families

The senior management perspective will look at the existing strategy and determine the impact on the projected financial performance, considering the growth, performance expectations, and market share. Management tends to have a higher-level view, reconciling both the short-term and long-term financials. Sales will have a more short-term view that is geared to quotas and meeting the current business plan. Sales will tend to focus more on customers and the detailed forecast by market, customer, or SKU. Marketing will have a more general view and focus on both the short-term view, looking at possible tactical exploitation opportunities, and the long-term view for the strategy for each product family. Therefore, it is essential that these different areas are represented in both the DDS&OP and AS&OP processes.

A very common way to look at various product families is through a portfolio analysis. Just as smart investors diversify their portfolios, a company needs to understand the diversification of its product portfolio. Typically, this portfolio consists of four distinct categories:

- **Existing products.** This is the current base of product offerings.
- **Line extensions and promos.** These are the current or potential permutations or derivations of the existing product offerings including promotional activity.
- **New to the company.** These are current and potential products that others produce but the company currently does not.
- **New to the world.** These are upcoming products that the company plans to produce that no other company has produced previously.

Where is the boundary between existing and new? Products are deemed new if the production and offering of the product could significantly disrupt the market and/or the supply chain or if the products need significant cross-functional resources to bring them to market. Products are not deemed new if a company already makes the product or if the company made the product in the past and still has the capability to make it today.

New products are those things that are new to the world or the company, additions to existing product lines, product improvements, process changes, or products that are being repositioned in the market.

There are the obvious kinds of new products and the AS&OP process needs to consider internal requirements for new things such as process development, infrastructure changes, sources of supply, or supply chain network reconfiguration. External new requirements could include changes in the market channels, government impact for regulatory compliance that varies by geography, or mergers and acquisitions.

The reality of new products is that it is more difficult to forecast demand and determine their impact on existing products. Additionally, there is usually an initial inventory investment required to have a proper product launch and the timing of the product introduction is usually critical. Having an effective stage and gate process is essential for managing the new product introduction cycle. Figure 6-4 is an example of a simple stage and gate process.

Many candidate products start in the investigation phase as the company considers the possibilities for desired future offerings according to their strategy. In Stage 1 the new product idea is assessed for overall fit into the strategy and possible bottom line contribution. Then the best ideas are considered in Stage 2 for a detailed business case. Considerations in Stage 2 include the impact of the new product on the constrained resources and how the contribution margin per unit of capacity constraint compares to the existing product portfolio. The surviving candidates then move to Stage 3 where the product goes into detailed product design and development. At this point a project team is assigned and resources are dedicated to the new product. In Stage 4 the final financial impact is determined in addition to the communication of requirements to the DDS&OP team. The DDS&OP team provides the feedback on the changes that must be made to the operating model, if any, and what impact those changes may have on other operating areas. In Stage 5, the product is launched and migrated to standard production processes within the DDOM.

Obviously, a company can have new or existing products, or some combination of the two, across its portfolio. Figure 6-5 depicts the most common portfolio patterns. Clearly, a company with revenue growth targets cannot achieve that if all product families are in the Stable

	Stage 1 Investigate	Stage 2 Business Case	Stage 3 Development	Stage 4 Test and Validate	Stage 5 Produce and Launch
Strategic Fit	Fit with strategy—yes/no	Strategic implications. Impact on business plan	On-going check—operating model aligned with strategy?	On-going check—operating model aligned with strategy?	On-going check—operating model aligned with strategy?
Marketing/Technical Activity	External assess market/customer/competition/technology	Detailed research major assumptions (i.e., market size, share, risk, etc.)	Customer/consumer testing, detailed design new product and process. Involve key suppliers	Brief sales and customers. Plan for launch	Manage launch activity. Move to "business as usual"
Production Activity	Broad feasibility—technology impact, capability/capacity implications w/current operating model	Understand new technologies, develop preferred processes, begin to develop remodel	Accelerate proto-types/samples, trials, experiments. Update operating model? Training	Refinements/improvements. Plan for ramp-up update operating model?	Ramp-up production. Move to "business as usual"
Financials	Rough-cut financial view, e.g., selling price range, price elasticity, cost profile/range, beneficial operation, contribution per unit of scarce resources	Research/refine data for decision making. Identify major assumptions, evaluate scenarios against operating model	Refine product costing, pricing strategy, capital and scale-up costs	Finalize product costing, launch price, new process costs and benefits, etc.	Fully integrated financials
Supply Chain Planning/Integration S&Op	Early visibility via new activities process and new activity review. Evaluate impact on current operating model	Include in integrated decision making (demand and supply)—assumptions/ sensitivity/ top down forecast (numbers an money)	Plan long timeline decisions—new suppliers/capital technology/material commitments, update operating model?	Develop detailed planning info/processes. Integrate top-down/ bottom-up views. Update operating model?	Fully integrated through integrated decision making and planning processes
Project Team Time/Resource	None—project team not appointed	Low (increasing)—developing business case and project plan	High—project activity and management	High—project activity and management	Medium (reducing)—"handing-over" to "business as usual"

FIGURE 6-4 Stage and Gate process

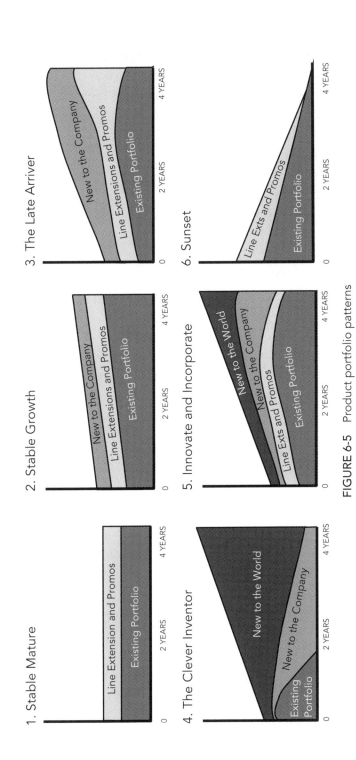

FIGURE 6-5 Product portfolio patterns

1. Stable Mature

Line Extension and Promos

Existing Portfolio

0   2 YEARS   4 YEARS

2. Stable Growth

New to the Company

Line Extensions and Promos

Existing Portfolio

0   2 YEARS   4 YEARS

3. The Late Arriver

New to the Company

Line Extensions and Promos

Existing Portfolio

0   2 YEARS   4 YEARS

4. The Clever Inventor

New to the World

New to the Company

Existing Portfolio

0   2 YEARS   4 YEARS

5. Innovate and Incorporate

New to the World

New to the Company

Line Exts and Promos

Existing Portfolio

0   2 YEARS   4 YEARS

6. Sunset

Line Exts and Promos

Existing Portfolio

0   2 YEARS   4 YEARS

Mature portfolio or in the Sunset Portfolio. This would indicate a major disconnect between the business plan and reality.

These six product portfolio configurations describe the most common configurations. The underlying company strategy is evident in the portfolio. For each of the portfolio types, there are four "R" questions that should be asked:

1. **Rate.** What are the expected future volumes of the portfolio?
2. **Resource.** What are the resources (including materials) required for the portfolio?
3. **Return.** How can the contribution margin relative to the required resources be maximized within the portfolio?
4. **Risk.** What are the risks involved with the portfolio?

The answers to the 4 R's will obviously differ by each different product portfolio and the company's position in the market. Using the 4 R's helps guide the conversation to discover underlying issues that need to be addressed in the Demand Driven Operating Model through the integrated reconciliation process. Let's examine these questions for each of the portfolio types in Figure 6-5.

- The Stable Mature portfolio expects current products to continue at a relatively stable rate of sales for the foreseeable future. An example might be products that are a necessary and specified component involved in a long-term contractual relationship.
  - **Rate.** Why is the product expected to stay stable?
  - **Resource.** Are the required resources expected to be stable (including suppliers)?
  - **Return.** Are pricing and direct material costs expected to also be stable?
  - **Risk.** What could jeopardize this stability?
- The Stable Growth portfolio expects the growth of these products to continue for the foreseeable future. An example might be a utility company in a growing municipality.

- **Rate.** What is the driver behind the growth? What is the potential impact from competition?
- **Resource.** How much can demand grow before additional resources are required?
- **Return.** Does pricing change as the growth continues? Do direct material costs remain the same as growth continues?
- **Risk.** What is the risk to the growth factor?

- The Late Arriver is willing to wait until another company incurs the bulk of the research and development costs and introduces a new product; it then develops a competitive product. The Late Arriver is not the first to market but can often become a market share leader by leap frogging the originator through enhancements or expanded features.
  - **Rate.** How to take market share? What promotions are required? How effective are the promotions?
  - **Resource.** Does this product affect other products? When is a step function of capacity addition required?
  - **Return.** Is there a change in supply cost as volumes increase?
  - **Risk.** Are we at risk for patent infringement? Are there others that will also replicate? How entrenched is the market leader?

- The Clever Inventor clearly expects their order winner to be breakthrough products that are new to the world and identifies them as the market leader. This company must invest in a significant research and development process to sustain their competitive advantage.
  - **Rate.** How quickly will the market adopt the innovation? How quickly can we develop and bring to market this new offering? How will the existing portfolio be impacted by the new products?
  - **Resource.** Do the new products use the same materials and resources?
  - **Return.** How quickly do the new products pay back the development investment? Is there patent protection available?

- **Risk.** Are there regulatory hurdles? What competitors could replicate and innovate on these offerings? What alternative technologies or offerings could halt market adoption?
- The Innovator and Incorporator is essentially a hybrid of the Late Arriver and the Clever Inventor. In some cases they come up with game changing technology and products and in other cases they launch competitive products that feature significant enhancements to a competitor's game changing product introduction. This profile is commonly seen in markets with two leading companies slugging it out for market supremacy often leap frogging each other with both innovations and enhancements. There is a heavy emphasis on rapid and long-term development.
  - **Rate.** How quickly will the market accept new innovations and/or incorporations? How quickly can we develop and bring to market this new offering? How will the existing portfolio be impacted by the new products?
  - **Resource.** Do the new products use the same materials and resources?
  - **Return.** How quickly do the new products pay back the development investment? Is there patent protection available? How long can these products maintain market leadership?
  - **Risk.** What legislation or government regulations could change? What new materials could be developed? What new technologies could affect the offerings? Are there patent implications for copying?
- The Sunset profile is composed of products that have a foreseeable end of life approaching regardless of enhancements. They have essentially become obsolete.
  - **Rate.** How quickly will demand deteriorate? Can we extend it through promotions?
  - **Resource.** What can be done with the resources that are available?
  - **Return.** Will margins be affected by volume?
  - **Risk.** What could accelerate the demand deterioration?

The important thing to remember in defining the portfolio and product families is the desired management information reports. Segmentation of products is necessary to allow relevant reporting in the integrated reconciliation process as well as the management business review. This process must be started with the end in mind. Analytics technology is available to provide multidimensional visibility, but management must decide what views will be relevant and what questions may be asked.

Step 2—Demand Plan

There are four elements for demand input into Adaptive S&OP: sales planning, demand creation, demand planning, and demand execution.

Sales Planning

Sales planning is the integration of sales knowledge and intelligence by account/geography/channel of specific market, customer, or competitor activities. In the strategic relevant range, that includes forecasting by product family and at times even specific SKUs when relevant within a family in the defined portfolio. When relevant, sales personnel should be encouraged and/or required to keep an updated "intelligence report" on all SKUs under their control. This intelligence report should be a living document to which Adaptive S&OP and DDS&OP have constant access.

Too often what they really know is hidden or obscured by simply aggregating to the family level. This issue is the challenge of simply using aggregation and disaggregation for forecast. The behavior of the family is not necessarily the simple sum of the SKU forecasts. Traditional formal planning prepares a planning bill of material with the expected percentage consumption for the product mix in the family. Frequently, the total percentage in the planning bill is greater than 100 percent in an attempt to cover the inherent variability of the actual demand. Extensive effort and expense is incurred in an attempt to get the mix percentages precisely correct. A key metric is how the actual demand tracks to that planning bill percentage. The inherent inaccuracy of the planning bill is then further exacerbated by the forecast-based MPS. This is an area where the DDAE model has a significant impact in eliminating this process. Inherent in the DDOM definition is the ability to plan and execute by ranges, which enables the strategy of being roughly right rather than precisely wrong.

Demand Creation

Demand creation is the proactive management of activities designed to significantly increase demand. Limitations to that demand are considered in the integrated reconciliation process. The demand plan is developed first using unconstrained possibilities. Demand planning is that process of forecasting all possible demand where demand execution is the interface with the markets and customer. This is the day-to-day management of actual demand through the Demand Driven Operating Model using the feedback loop from DDS&OP. Demand execution assesses outliers and exceptions, looking for an early warning of a significant market shift.

Demand Planning

The demand planning review is the process to agree to the future unconstrained demand plan. Forecasting is necessary and relevant for the strategic relevant range. Documenting assumptions behind the forecast is critical to establish at least some level of trust in the forecast numbers. In this way different views and scenarios are actively sought, reconciled, and compared to the current Demand Driven Operating Model capability. This is another reason why the sales market intelligence report must be accessible throughout the planning process. Bidirectional reconciliation is accomplished in the DDS&OP process.

In addition to product family portfolio performance in the general market, most products go through a product life cycle of launch, growth, maturity, decline, and exit as shown in Figure 6-6. The demand plan considers where the different product families are in that product life cycle and decisions are made to extend the product life or manage end of life. Note how the forecast range also changes as the product life cycle progresses, suggesting that the operating model must also adapt to accommodate this required range.

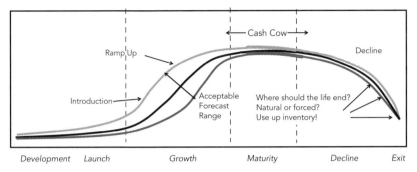

FIGURE 6-6 Product life cycle strategies and forecast range

The introduction of a replacement product also impacts the decision-making process. The end of life for the old product must be managed at the same time as the product introduction. These ramp up and ramp down buffer configurations are part of the transition process that DDS&OP manages. As the product demand continues to decline, the consideration of flow-based opportunity cost changes may necessitate the forced elimination of the product sooner than expected. Are these products consuming scarce material, capacity and/or space that other products could better utilize?

Demand Execution

Demand execution is accomplishing those activities that enable the demand plan. This includes activities in the tactical relevant range to help make the demand plan become a reality. For example, if the actual sales are lagging, then pricing and promotion adjustments are considered. Managing exceptional demand, outliers in the positive or negative direction, is also accomplished by demand execution. DDS&OP plays a prominent role in demand execution within the tactical relevant range.

Forecasts Ranges and Forecast Error Ranges

Ian Wilson, a former student of Dick Ling, once said, "however good our research may be, we shall never escape from the ultimate dilemma that all our knowledge is about the past and all our decisions are about the future." This is a critical point to remember. The best information there is about the future is a forecast and while forecasts always contain error, having an operational model that can perform with agility within a defined range allows a company to leverage those forecasts more effectively. The penalty for error is significantly reduced because the DDOM reacts only to actual demand and has built in buffer ranges to absorb inevitable problems with forecast variance. These different views and scenarios are actively sought, reconciled, and compared to the current operating model capability and is the strategic recommendation and validation loops from DDS&OP.

Management of the demand plan should be thought of as a range rather than just a single number. A relevant forecast is an expectation with a range of potential error. This range is reconciled by the Adaptive S&OP team for adherence to the business requirements and validated by the DDS&OP team for operational capability.

A range of possibilities is more realistic than trying to always hit a series of single numbers. In reality, however, any single number comes with a range in the form of forecast error. One of the advantages of the DDOM is that any single number automatically translates to a range of potential as every type of buffer has an easily defined range in which it can tolerate increases or decreases of actual demand. The configuration of the DDOM (including buffer sizing) must address this error range and provide mitigation for the variability through its buffers. If this range of high and low (forecast error range) can be approximated, then it is a relatively simple step for DDS&OP to compare that range with the defined buffer tolerances in the model. If the higher rate of demand materializes, will the buffers be able to absorb that before the model is reconfigured? If the lower rate of demand materializes, how exposed is the DDOM from a working capital perspective before it is reconfigured?

Step 3—Supply Plan

This step is fundamentally asking the question, "What capabilities does the DDOM have in the strategic time frame?"

The capacity strategy needs to complement the demand strategy. Capacity cannot typically be added proportionately to demand (of course except for short-term overtime). Buffers in the DDOM, however, can be placed and configured and reconfigured in proportion to demand (both actual and anticipated). Buffering is not adding capacity, it is essentially realigning capacity to be better utilized in the face of a higher volume or variability in demand. But the placement, configuration, and reconfiguration of the buffers can only go so far before the system output becomes constrained and additional overall capacity is needed.

Capacity tends to be added in step functions with an impact on the overall invested capital, fixed expense change, and lead time to consider for deployment. This is one reason why the Adaptive S&OP process needs to look out into the future as far as necessary to identify these capital investment requirements. As described earlier, ROI is the primary metric to guide decisions as discussed in the DDS&OP chapter concerning the contribution margin per unit of constrained capacity.

The supply review takes the consolidated unconstrained demand plan and new activity to determine the feasibility of that plan given internal and external supply capability. The output of the supply plan is an achievable supply plan by product family in volume, timing, and cost. The supply review provides the strategic settings for the DDS&OP process to define the Demand Driven Operating Model. Therefore, the current DDOM design and general capability must be understood by the Adaptive S&OP team. This is not a triviality!

The supply plan manages the high-level allocation of capacity consumption. For example, if the company strategy is to be a leader in product development, then sufficient capacity must be allocated for R&D. If the company strategy is quick response to a wide variety of products, then buffer capacity must be available to mitigate that demand variability. Special focus is on any critical supply resource (internal or external) that could constrain supply to ensure that the highest contribution margin per unit of that capacity is allocated first for production.

In this supply planning process, the role of key suppliers is a priority. Suppliers are a critical part of the overall supply plan because the limiting capacity constraint may not be within the company's four walls but rather in the supply base. Consider which suppliers are providing significant capacity to support the company's activities. Supplier location and performance are critical factors to consider in a supply risk plan. Suppliers that have long lead times and high variability mandate a different DDOM configuration than a supply base that has short lead time and low variability. The company must recognize the importance and risk from the supply base.

A company that has an effective Adaptive S&OP process will also include key suppliers as part of the overall Adaptive S&OP process (see Stage 5 of the DDAE development stages in Chapter 8). This recognizes the importance of that supplier and gains additional support from the supplier. The supplier can be part of the team, which means that change, collaboration, and overall supply chain flow will be managed more effectively. The reward that the company gets for including suppliers is usually shorter lead times as suppliers are no longer second guessing the company's needs. The benefit for the supplier is a better understanding of demand input into their own S&OP (or Adaptive S&OP) process.

In the supply plan process, the capability risk and/or gaps are identified. The process is done by manufacturing product family and the capacity requirement is consolidated to the critical resource capacity to determine if resource changes, investments, or process changes are needed. Figure 6-7 shows the relationship from the strategic definition of product families to the tactical consolidation that checks for a rough-cut fit of demand to capacity. The DDOM then utilizes the buffer profile and the average daily usage (ADU) by SKU for relevant parts. Note that these

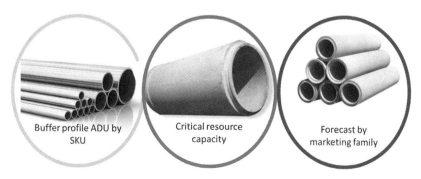

Buffer profile ADU by SKU

Critical resource capacity

Forecast by marketing family

FIGURE 6-7   Rough cut capacity planning

capacity planning perspectives are not directly connected. Each process is decoupled from the next but reconciled in a bidirectional closed loop process. The traditional process of aggregation and disaggregation is replaced by this bidirectional reconciliation and adaptive operating model capability. Relevant information is considered in each of the different time frames. Roughly right is much better than precisely wrong.

Step 4—Financial Plan

The financial planning process examines the projected financial performance and determines what factors are relevant over the strategic time frame. In the financial plan, the product portfolio is rated against its relative market share and how it is generating cash as compared to market growth rate and the usage of cash. This segmentation is another application of the product family definition process to group products into financial families for better focus and easier management. This segmentation typically has four quadrants: high growth market products that generate a high level of cash, high growth market products that generate a low level of cash, low growth market products that generate a low level of cash, and low growth market products that generate a high level of cash. Just as the marketing and manufacturing families have different strategies based on the position of that product family in the product life cycle, the financial strategy varies by the product family's position in this financial grid. High growth/high cash products are usually the new products and are also subject to higher volatility that must be accounted for in the buffering strategy of the operating model. The low growth/low cash products may only make sense to maintain in the portfolio to cover fixed expenses because there is no higher contribution margin product available.

A financial family that is in a slower growth market but generates a high level of cash needs to be protected from disruption. These products are usually referred to as "cash cows" and represent a strong financial contribution for the company. When evaluating new products for introduction, the desire is usually to minimize the impact on these high cash generation products. These products customarily are high volume products that have fewer problems in the manufacturing process.

A common mistake made in many companies is allocating overhead and other fixed costs based on labor hours. This is the classic mistake of mixing relevant and distortive information. This process of fully loading fixed and variable costs to a per unit part cost can mask the discovery of these cash cows by adding these non-relevant costs. When calculating part costs, the only relevant costs are the truly variable costs. The difference between the selling price and this variable part cost is the contribution margin, which measures the rate of cash contribution. The relevant metric for comparing financial performance is then the contribution margin realized per unit of capacity limitation. Many companies have killed off their cash cows because of the financial practice in the company of evaluating products for rationalization based on a total loaded cost subtracted from revenue per part. Only when the overall ROI of the company starts to decline is this strategic error discovered—and many times it is too late.

In addition to managing the individual financial strategy by financial product family, general financial considerations include how the business plan is tracking to actual performance as well as the overall working capital availability. The financial plan brings to the plan the constraints imposed by financial reality like cash flow, shareholder expectations, and bank covenants. A company may forego some operational improvement because of the impact on the financial statements. For example, dropping inventory too quickly can make a company look unprofitable on the P&L statement. If that improvement causes the GAAP financial statements to run afoul of bank covenants, the rate of improvement may need to be slowed. This override would then be reflected in the Demand Driven Operating Model and how the buffers are sized. Typically, this is not an issue in privately held companies because the ownership of the company is also involved in the management of the operational model. When management and ownership are decoupled, expectations must be carefully managed and/or the rate of improvement may need to be slowed to allow the financial statements to "catch up."

The DDS&OP process provides the forward-looking operating model expectations that then can be translated into the financial plan and reconciled with the shareholder's financial expectations. The financial plan considers the ROI by product family and examines relevant metrics like total sales per month, contribution margin/month, and actual contribution margin per time unit on critical resources. These items can change as suggestions are made to address capacity issues, such as how adding a third shift changes fixed cost. Re-engineering a product for lower direct material cost will change long-range variable costs. The financial plan establishes pricing strategies in concert with the Sales and Marketing department including the impact of the price elasticity of demand. Implementing a promotion to fill capacity may result in an unintended negative consequence financially. This is a reason why a representative from finance needs to be on the DDS&OP team as well as the Adaptive S&OP team.

The financial representative that participates in the Adaptive S&OP team must be well versed in both managerial and cost accounting methodologies to understand the relevant financial and reporting impact of the different scenarios that are considered.

Step 5—Integrated Reconciliation

A critical role in the Adaptive S&OP process is the integrated reconciliation process. This need was identified several years ago during the evolution of Breakthrough S&OP.[2] The integrated reconciliation process is where the portfolio planning process, demand plan, supply plan, and financial plan come together. Note that in Figure 6-2 these planning processes do not drive each other in a linear fashion but rather each of the four planning processes affect the other three in a holistic manner. The Adaptive S&OP team must have someone involved in each of the other processes, so that there is an awareness that there will be issues that need to be escalated and reconciled. Additionally, an effective and efficient integrated reconciliation process is heavily reliant on the quality of the documented assumptions.

The Adaptive S&OP team is not a super team that makes all the decisions, and integrated reconciliation is really only the first step to create or amend the Business Plan. Line management still needs to be where most decisions are made, and accountability established. Operational flow-based metrics align the line management to the overall corporate objectives. At this step the Adaptive S&OP process enables management to reach a consensus, make decisions, identify misalignments and gaps, and make recommendations on a path forward. Typically, this process includes finance, new product development, sales, marketing, supply chain management, and operations representatives. Participation does not have to be full time; it is more important to have the right people involved. People who have an active role in an area are usually the right people. Temporary participants may be added due to a specific situation or business condition that arises. A team member may wear more than one hat. This process supports the development of the future leaders of the company and is not merely a structure for Strategic planning. This is a crucial aspect of a mature Demand Driven Adaptive Enterprise. Very experienced people are needed. Serving on this team is recognition that the participants are being groomed for higher positions in the company. Communication skills are important in addition to analytical and computer skills.

At the beginning of the development of the process, there is an initial meeting of the Adaptive S&OP team with the Executive team to gain understanding of strategy and direction, including the dimensions of the go-to market strategy, operational strategy, financial strategy, and innovation strategy. An information process map is also defined for the decision-making process. This defines the location where decisions need to be made and what empowerment can be taken by the team. This is the "thoughtware" employed in the organization and will vary by type of decision.

The integrated reconciliation process helps facilitate and ensure that the necessary decisions for all areas of the business are made in an agile way that supports the strategic direction and financial goals of the company. This process adapts the plan to relevant changes as quickly as possible and facilitates communications, decision making, and alignment in a cross-functional approach to operating the business. It is important to realize that bidirectional communication is required. The integrated reconciliation process forces a cross-functional way of operating and this requires a collaborative process.

The integrated reconciliation process considers strategic scenarios like new opportunities in the market place, new product timing, deviations from the business plan, impact of supply problems, evaluation of supply chain risks, or any issue emerging from the monthly planning process. Executive management defines the expectations for the Adaptive S&OP process. This includes the development of an understanding of relevant ranges and the information specific to those ranges. To gain momentum, the process is driven by decisions needing implementation. These decisions can include changes in the priority of customers, regions, products, etc. These decisions then shape the future company strategy and the necessary changes to the Demand Driven Operating Model. The four perspectives (supply, demand, financial, product portfolio) come together in this process.

Integrated reconciliation is all about a spirit of and mechanism for collaboration and alignment in the company. This process builds trust across the team. It is critical to understand that many decisions need to be made by Operations, Sales, Marketing, and Development leaders and not by the Adaptive S&OP team. The Adaptive S&OP team's role is to facilitate the change process and ensure alignment by identifying the variance between the intended direction and actual results. This includes resolving issues, getting decisions made, and seizing on potential opportunities defined in the DDS&OP process. For example, in Chapter 5, the issue was raised by the DDS&OP team about the economic feasibility and operational impact of a preliminary international expansion.

This Adaptive S&OP process is about managing and embracing change in an agile yet aligned fashion. The integrated reconciliation process is where the people side of "sense, adapt, and innovate" is centered and facilitated. This integration process helps management focus on what is changing through the scenario evaluation process. There are several questions behind the scenarios. What will probably change and do we understand and agree on the assumptions behind the change? Do we understand the impact of the changes? Finally, what decisions have already been made or what decisions still need to be made? Different triggers can cause a change in planning. These can be an event or series of events or just be a result of the re-planning process itself.

The Adaptive S&OP team has the responsibility of developing material for executive management to make informed decisions. The leader of this team must be at all executive meetings in order to understand the current direction and priorities, and answer any questions about the team's recommendation. The leader will then initiate action by the team to satisfy decisions made by management.

This process is critical to having a management team capable of leveraging and exploiting the possibilities for the strategic plan. This process will evolve and change with experience. Integrated reconciliation is learned by doing and this should be the goal of all members of the team and Executive Management.

Step 6—Demand Driven S&OP

DDS&OP is more fully described in Chapter 5, so this section will summarize the role of DDS&OP in the AS&OP process. DDS&OP provides the "what if" scenario evaluation capability with the resulting analysis in the AS&OP process. Suggested changes in new product introductions, seasonal fluctuations, major promotions, changes in capacity, and anticipated changes in demand are all evaluated against the current DDOM to determine what input parameters must be changed. Each scenario evaluated provides a strategic, financial, market, and operational perspective. From the strategic perspective, information includes the impact on the business plan and identification of risk. From the financial perspective, the scenario provides insight into profitability, working capital, and overall ROI. From the market perspective, the scenario establishes the impact to market share and if the product moves to a different financial family. From the operational perspective, the feasibility of the plan is established and compared to what is already in place. The past focus of scenario planning has been on resource optimization. Now it is possible to focus on maximizing business results directly.

As stated previously, the Master Production Schedule (MPS) is gone from the S&OP process. No longer is there a need for a schedule by SKU by date to drive the formal planning system. That practice ensures that the company will have too much of the wrong inventory while experiencing shortages. Now planning is accomplished by the strategic positioning of decoupling and control points protected by time, inventory, or capacity buffers activated by real demand. Resources are activated based on what can and will be sold rather than a detailed schedule that is precisely wrong. However, the function of tactical planning still exists. Many companies today try to use the MPS for tactical planning. Tactical planning now determines the capability of operations to manufacture what is being sold and the tolerance of fluctuations that can be handled. DDS&OP is the missing link in the overall S&OP process. Now it is possible to have a bidirectional tactical reconciliation hub that allows a company to sense changes in market demand and adapt planning and production while pulling from suppliers in real time.

Figure 6-8 is about managing choices through the range that was established through this bidirectional management process. The demand plan manages the choices within the strategic parameters. Performance is never a predictable straight line but rather a series of choices that navigates the best route within the strategic boundaries. Those boundaries are easily defined by what is possible with the market and the DDOM capability.

One of the items that causes re-planning could be the movement of the product through its life cycle, and it requires special consideration. The Adaptive S&OP team helps with scenario planning and evaluation of proposed business plans against the necessary operational capability as part of the DDS&OP process. An example was described in Chapter 5. This process ensures that the assumptions supporting the numbers are documented and communicated to executive management. This bidirectional integrated reconciliation process helps assess financial impact of changes and communicates necessary positioning changes to

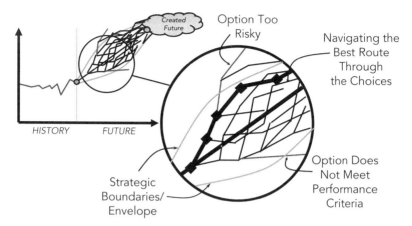

FIGURE 6-8 Navigating the demand plan

the Demand Driven Operating Model. This includes establishing inventory decoupling points based on:

- Customer Tolerance time
- Market Potential Lead time
- Sales Order Visibility Horizon
- External Variability
- Inventory Leverage and Flexibility
- Critical Operation Protection

The establishment of decoupling points is a strategic decision that determines customer lead time and working capital investment. Additional control points are then added based on pacing resources, exit and entry points to the system, common points (convergence or divergence), or points that have notorious process instability. The bidirectional tactical reconciliation accomplished by the DDS&OP process then communicates necessary buffer sizing changes to the Demand Driven Operating Model. These buffers can be inventory, time, or capacity. The operating model is affected by significant planned events such as new product build-up, plant shut down, or seasonality with planned adjustment factors.

The DDS&OP process also provides feedback from the Demand Driven Operating Model with respect to the operating model reliability, stability, and speed/velocity. See Chapter 4 for more detail on the DDOM. This process is leveraged by the Demand Driven Adaptive Enterprise Model as described in Chapter 3 to provide a creative adaptation to satisfy current customers, build the marketplace, and improve company ROI.

System return can be maximized according to the relevant model factors for volume and rate. This process also points out and prioritizes lost ROI opportunities based on the current operating model performance. New business plan scenarios can then be objectively evaluated based on the current operating model and how it is performing for working capital, space requirements, capacity, or other relevant metrics. This was described in Chapter 5 using a DDS&OP example.

Step 7—Management Review

The management review is the senior level review of the latest plan and makes the final decision on its validity. Only those items requiring senior management's attention should be escalated to this level. In this review the focus is on how the company is performing against key strategic metrics and whether the current performance is supporting the business plan. The review develops the future direction for the company and identifies what vulnerability and opportunities now exist and how that has changed over time. The decisions that have been escalated from the Adaptive S&OP team need to be made in a timely manner to ensure that both the Adaptive S&OP and DDS&OP teams have the information necessary to execute the strategy. In a small or medium sized company, the Adaptive S&OP and executive teams are most likely the same people, allowing for a quicker pace.

Preparing information for the review includes ongoing planning assumptions, major assumptions or changes in this cycle, identification of emerging issues, decisions made in previous steps, and identifications of risk and opportunities, in addition to identifying the decisions that are required by the executive team. The scenario evaluation process is essential to understanding which decisions are key to escalate to the management review level.

There are three types of information for the management review: data, graphs, and text as shown in Figure 6-9. The data (numbers) alone are not enough. The graphs help to understand the numbers in a highly visual way. The underlying assumptions explain and create trust in the numbers and develops trust in the people who are providing those numbers. The assumptions support the data and help with understanding and agreement. Developing trust is critical for successful management teams.

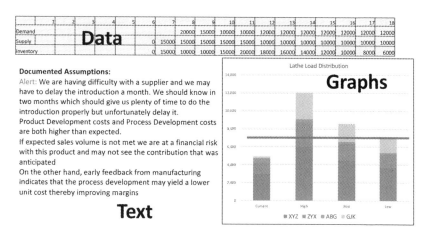

FIGURE 6-9   Management review supporting documentation example

Strategic Metrics

Adaptive S&OP is responsible for measuring the overall performance of the company. Strategic metrics emphasize contribution margin, working capital, and how the company is performing in the marketplace. Figure 6-10 displays the strategic metric objectives and messages.

Metric Objectives	The Message Behind the Objective
Contribution Margin (cash generation rate)	Drive innovation (internal and external) and growth to increase cash generation capability (RATE)
Working Capital (inventory and cash and credit)	Ensure proper levels of working capital to protect and promote flow in the short and long term
Customer Base (market share, sales and service and quality)	Ensure and grow a solid base of business for the enterprise (VOLUME)

FIGURE 6-10 Strategic metrics objectives and messages

With Figure 6-10, we have come full circle to Chapter 1; it shows the three vital metrics that determine an organization's success and sustainability. Compromising any one of these critical areas too far can push the organization into collapse.

Contribution margin is the rate at which the organization is capable of generating cash. The message behind this objective is for the company to drive innovation toward increasing cash generation capability. This generation and the innovations driven must be viewed in context of key factors with regard to the market, materials, and DDOM capability. Are there market segmentation possibilities to allow for higher pricing? Are there design changes that can increase the rate? Are there alternative sources of supply that alter the rate? Are there process or capability changes that accelerate the rate?

Working capital includes inventory, cash, and credit. Too often inventory is labeled as "bad" and earmarked for elimination. Plainly stated, inventory is an asset and should be expected to deliver a return on that investment just like any other asset in the company. Too little inventory and the company cannot react to demand pulses or unique and lucrative short-term opportunities. In a more volatile world, there will be more short-term opportunities! Too much inventory and the company is not leveraging its capability.

The same phenomenon is seen for cash and credit. Too much cash and the company is a target for a hostile takeover or acquisition. Too little cash and the company cannot pay its bills, which leads to suppliers not delivering necessary components. Too much credit and the company is swimming in debt and loses a great deal of profit to servicing that debt. Too little credit and the company cannot pounce on last-minute opportunities because there is no flexibility.

Customer base is about managing strategy for volume today and tomorrow. It means the ability to service current demand at a high level and drive market innovation for the future. A company can be highly efficient with wonderful products and services, but if these don't meet a need in the market the company will go bankrupt. Especially in the recent days of social media, monitoring and maintaining the company's reputation and quality of the product is essential. How a complaint or bad service is handled becomes a very public demonstration for the company.

Summary

Adaptive S&OP is a necessary condition for a mature Demand Driven Adaptive Enterprise. The company must have a process that maintains coherence between subsystem behavior and overall system strategy for the longer range: the strategic adaptive cycle. AS&OP translates the overall strategy into a realistic business plan and direction for DDS&OP to put into execution. This integrated business planning process engages all areas of the business with cross-functional communication that yields more relevant information for all participants. It ensures that all levels of the organization understand the aggregate and detailed goals of the organization and are committed to executing them. In short, the AS&OP process ensures that the business plan is complete and attainable on a continuous basis and deals with change to make the company adaptive. Adaptive S&OP is all about people and leadership, and of course, flow.

In summary, the combination of DDS&OP and Adaptive S&OP brings Dick Ling's original vision for S&OP to reality. Sales and Operations planning is the integrated business process that provides management the ability to strategically direct its businesses to achieve competitive advantage on a continuous basis by the protection and promotion of return on investment. Product innovation, customer focused marketing plans for new and existing products, operations strategy, and the financial strategy are managed on a continuous basis by an integrated reconciliation team to enable the company to sense, adapt, and innovate successfully across the supply chain.

CHAPTER 7

Closing the Loop
with the DDAE Model

Now that the three components of the DDAE Model have been described in some detail, we are ready to circle back to some of our basic requirements mentioned in the first two chapters of this book. We will start with the four prerequisites to relevant information.

A Model Built for Relevant Information

The DDAE model was designed specifically with the prerequisites for relevant information as discussed in Chapter 2.

Prerequisite #1: Relevant Ranges

The DDAE model uses three connected and reconciled relevant ranges: Operational, Tactical, and Strategic. Figure 7-1 illustrates the three relevant ranges, including their typical definitions in the Demand Driven Adaptive Enterprise Model.

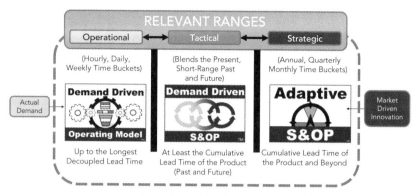

FIGURE 7-1   DDAE model relevant ranges definitions

The Operational Relevant Range is typically expressed in hourly, daily, and weekly buckets but spans up to the longest decoupled lead time of a part/SKU. Decoupled lead time was introduced in Chapter 4 and is an important aspect of configuration, buffer sizing, and replenishment planning in the Demand Driven Operating Model. This is the time range that matters between decoupling points (which defines planning horizons in DDMRP and the Demand Driven Operating Model). Day-to-day operations are managed within the Operational Relevant Range.

The Tactical Relevant Range combines the present with the short-range past and near-term future. It typically spans at least the cumulative lead time into the past and into the future. Thus, it typically represents a total period (present, future, and past) at least twice the cumulative lead time. The Operational and Strategic Relevant Ranges are reconciled in the Tactical Relevant Range through the DDS&OP process explained in Chapter 5.

The Strategic Relevant Range is typically expressed in monthly, quarterly, and annual time buckets. This range typically starts at the product cumulative lead time in the future and looks further into the future depending on the time required to adapt the model for capability and capacity. Business Plan parameters are devised and revised within the Strategic Relevant Range through the Adaptive S&OP process explained in Chapter 6.

Prerequisite #2: Tactical Reconciliation
Between Relevant Ranges

The DDAE Model uses a process known as Demand Driven S&OP to reconcile the strategic and operational relevant range explained in Chapter 5. Whereas convention tends to be a top-down driven linear process culminating in a constantly changing Master Production Schedule, DDS&OP uses a constant bidirectional iterative process to adapt to required changes driven from past performance and known or planned future events.

Ultimately, tactical reconciliation isn't about precise targets, schedules, or plans; it is about the management of various ranges for capability. These ranges are defined in the configuration of the DDOM and carefully monitored and managed by the DDS&OP team. Actions are driven around how the model performs within these ranges and/or how it is expected to perform in the near-term future.

The whole DDAE Model is driven around a basic philosophy: to be approximately right, not precisely wrong. Convention uses specific numbers being fed to a Master Production Schedule that is then fed to MRP, which directly launches supply orders with specific timing against them. These specific numbers are guaranteed to be wrong. The DDAE model still uses specific numbers representing predictions, but those numbers then automatically produce ranges of capability in the DDOM. Those specific ranges can be changed either by changing the input number or by changing things like stock and time buffer profiles. It is DDS&OP's job to manage the DDOM to those ranges. The specific numbers will be wrong and will vary within a range, but the ranges will be approximately right and will self-correct or adjust to get even more right! The DDS&OP team has the most important role in constantly reinforcing this concept up and down the organization. Management of the business moves from transactional control to process and/or model control.

Figure 7-2 depicts how the three components of the DDAE model work together over these three relevant ranges. DDS&OP is managing the past, present, and tactical future while the DDOM is managing the operational range (present). Adaptive S&OP is adapting the picture of the strategic relevant range as information becomes available, providing a range of possibilities and evaluating scenarios for selection by management for the future strategic direction.

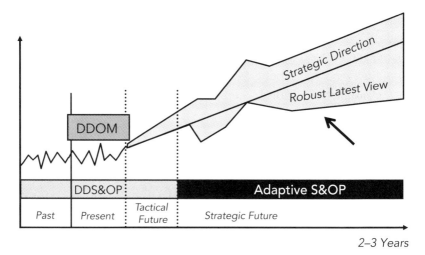

FIGURE 7-2 DDAE Model components and relevant ranges

Prerequisite #3: A Flow-Based Operating Model

As described in Chapter 4, the DDAE Model incorporates a flow-based operating approach called the Demand Driven Operating Model (DDOM). A Demand Driven Operating Model (DDOM) is a supply order generation, operational scheduling, and execution model utilizing actual demand in combination with strategic decoupling and control points with stock, time, and capacity buffers in order to create a predictable and agile system. It promotes and protects the flow of relevant information and materials within the operational relevant range. As described in Chapter 5, a DDOM's key parameters are set through the Demand Driven Sales & Operations Planning (DDS&OP) process to meet the stated business and market objectives while minimizing working capital and expedite-related expenses.

The DDOM is designed around four basic elements:

- **Pacing to actual demand.** The Demand Driven Operating Model uses only actual demand for supply order generation. There are no planned orders and no Master Production Schedule used in the DDOM.

- **Strategic decoupling points.** The Demand Driven Operating Model uses strategically placed decoupling points to compress lead times, shorten planning horizons, and dampen demand and supply variability simultaneously.
- **Strategic control points.** The Demand Driven Operating Model uses strategically placed control points for scheduling in addition to resource and order synchronization.
- **Dynamic buffering.** The Demand Driven Operating Model protects its decoupling and control points through dynamic stock, time, and capacity buffers.

The heart of the DDOM is the innovative method of supply order generation and execution known as Demand Driven Material Requirements Planning (DDMRP). DDMRP utilizes strategically determined decoupling point buffers to compress lead times and minimize the distortion to relevant information (transfer and amplification of demand signal distortion) up the supply chain and the distortion to relevant materials (supply continuity variability) down the supply chain. Detailed resource scheduling is accomplished through Demand Driven Scheduling. Demand Driven Scheduling utilizes the strategically determined placement and scheduling of control points protected by a combination of stock, time, and capacity buffers. Demand Driven Execution is the management of open supply orders and released manufacturing orders against the real-time status of stock, time, and capacity buffers.

Prerequisite #4: A Flow-Based Metrics Suite

The DDAE model uses different metric emphases within each of the operational, tactical, and strategic relevant ranges to promote and protect flow today and into the future. Operational metrics (described in Chapter 4) emphasize reliability, stability, and velocity to determine relevant information and materials in the Operational Relevant Range. Tactical metrics (described in Chapter 5) emphasize system improvement, waste reduction, operating expense control, short-range cash contribution, and additional potential (volume and rate) in order to determine relevant information and materials in the Tactical Relevant Range. Strategic metrics (described in Chapters 1 and 6) emphasize contribution margin, working capital control, and customer base control and development.

	Metric Objectives	The Message Behind the Objective
Operational	Operational reliability	Execute to the model, plan, schedule, and market expectation
	Operational stability	Pass on as little variation as possible
	Operational speed/velocity	Pass the right work on as fast as possible
Tactical	Operational improvement and waste reduction (opportunity $)	Identify and prioritize obstacles and/or conflicts to flow
	Operating expense control	Spend minimization to meet the requirements of the market and the DDOM design
	Operational strategic contribution	Maximize system return according to relevant model factors and tactical opportunities (volume and rate)
Strategic	Contribution margin (cash generation rate)	Drive innovation (internal and external) and growth to increase cash generation capability (RATE)
	Working capital (inventory and cash and credit)	Ensure proper levels of working capital to protect and promote flow in the short and long term
	Customer base (market share, sales and service and quality)	Ensure and grow a solid base of business for the enterprise (VOLUME)

FIGURE 7-3   The DDAE flow-based metric suite

Figure 7-3 is a composite picture of the flow-based metrics in each of the relevant ranges seen in Chapters 4, 5, and 6. In each case there is a metric objective and a message behind the metric. This allows companies to build business specific metrics that fit these objectives and messages within their unique Demand Driven Adaptive Enterprise model and market circumstances. Since all of the metric objectives in each relevant range are designed specifically for flow, the relevant ranges directly connect to each other through the iterative connections between the three components of the DDAE model.

A Model Built for Complex Adaptive Systems

Now let's close the loop on specifically summarizing the DDAE characteristics and how they align with complex adaptive systems characteristics.

The DDAE model is designed specifically for the complex adaptive systems (CAS) described in Chapter 1. The DDAE model incorporates a framework for CAS characteristics to be better identified, focused, and utilized. As described in Chapter 1, CAS are constantly evolving through a cycle of emergence, feedback, and selection where:

- Emergence is a reconfiguration of the system triggered externally or internally.
- Feedback is a set of defined signals and triggers that are monitored by adaptive agents.
- Selection is decisions, actions, and learning in response to the signals and triggers, which may or may not result in another reconfiguration.

As first mentioned in Chapter 3, the DDAE model uses two distinct adaptive loops or cycles to drive adaptation. These cycles are different because they operate in two different relevant ranges. The two cycles do, however, meet in the middle of the model with Demand Driven S&OP described in Chapter 5. Figure 3-2 depicts the two adaptive loops in the DDAE Model.

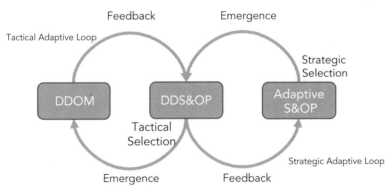

FIGURE 3-2 The adaptive loops of the DDAE Model

Figure 7-4 is the CAS characteristics and explanations from Chapter 1 but also contains the DDAE model attributes for each of the CAS characteristics. All of these attributes were described in Chapters 4 through 6.

CAS Key Characteristic	Explanation	DDAE Model Attribute(s)
Boundaries	All system boundaries are defined by their subsystems, hierarchies, adaptive agents and signal sets to trigger action and interactions between them. No true boundary exists in a CAS because they are always part of a larger ecosystem but boundaries are practically defined by the limits at which the system's adaptive agents can act and affect change within.	The DDAE model uses specific boundaries to help direct adaptive agents. These boundaries are best summarized in the distinction between relevant ranges primarily defined by decoupled lead times and cumulative lead times.
Coherence	A CAS depends on its subsystems to align their purpose and actions with that of the system. Subsystems that are not in alignment can cause coherence to break down and push the system into chaos.	The DDAE model is built around the concept of the flow of relevant information and materials. Metrics are designed specifically to reinforce this conceptual basis at all levels and relevant ranges.
The Edge of Chaos	All CAS operate in a zone between stable equilibrium points and total randomness otherwise known as chaos. These zones have been learned or defined and are monitored by adaptive agents through sets of signals. Adaptive agents will act and even self-limit in order to attempt to keep the system out of chaos.	The DDAE model uses defined ranges and metrics that carefully monitored by relevant personnel in order to keep the system out of chaos. Examples are stock, time, and capacity buffer monitoring.
Self-Organization, Innovation and Emergence	CAS have emergent properties or events in which adaptive agents self-organize based on signal strength in order to solve an issue. Through feedback and selection they will learn and innovate in order to bring the system to an evolved state.	The DDAE model is designed to highlight system characteristics that can lead to emergence or innovation. Operationally, this emergence or innovation is typically driven by the DDS&OP process. Strategically, it is driven by Adaptive S&OP.
Signals	All CAS use defined signals to communicate with and within subsystems as well as with other systems that they interact with. System coherence and signal alignment is the primary consideration in determining the actual signals, their triggers, and subsequent instruction sets.	The DDAE model uses highly visible and intuitive signals in order to direct personnel at all levels to maintain coherence and or keep the system out of chaos. Examples of these signals were featured in Chapters 4 and 5.
Adaptive Agents	All CAS are dependent on adaptive agents to receive, interpret, and react to signals. They are adaptive because they are responsible for identifying and learning about changes in the patterns of signals and making recommendations for changes and innovation.	Adaptive agents are present in all relevant ranges of the DDAE model. Adaptive agents in operational range would be buffer managers that attempt to protect schedules. The DDS&OP team are the adaptive agents in the tactical range. The Adaptive S&OP team are the adaptive agents in the strategic range.

CAS Key Characteristic	Explanation	DDAE Model Attribute(s)
Signal Strength	All CAS depend on adaptive agents to recognize and prioritize across signals with system coherence always in mind.	The DDAE model uses highly visible and intuitive signals that are designed to quickly convey a sense of priority to maintain coherence in all ranges.
Feedback Loops	All CAS depend on feedback loops to provide adaptive agents the ability to measure, sense, and adapt subsystem and system performance in order to maintain coherence and drive innovation and emergence.	A significant feature of the DDAE Mmodel is the feedback loops designed to drive the tactical and strategic adaptive cycles. Examples of these are best seen in Chapter 5.
Resilience	Resilience is the ability of a CAS to return to equilibrium after a large disturbance (either imposed or self-imposed). If a CAS is too rigid the disturbance will push it into chaos.	The placement, sizing, and management of stock, time, and capacity buffers are a prominent feature of the DDAE model. While no system can protect against every potential disturbance DDAE models are much more resilient than the conventional MPS-MRP driven model.

FIGURE 7-4   CAS-related DDAE Model attributes

Demand Driven Project Management

How do we bring our strategy to life? The enablement of our strategy is typically through the completion of projects. These projects are to support and/or enhance the company's offerings to the market in the future (market driven innovation). These types of projects might include new product introductions, the construction of new facilities, or the research and development of a game changing technology. These projects reside in the strategic relevant range. We will call these projects "Market Driven Innovation" projects. These projects are key components of a company's strategy over the long-range future; they bring market driven innovation to fruition. An example might be the development of a new type of boring technology for a large construction equipment manufacturer.

In Market Driven Innovation projects, the flow of information and integration has a target: to create something of value that allows the company more opportunity, flexibility, or capability within a certain (often critical) amount of time. Often, if the timing is late, the value of the project's target is diminished or, in extreme cases, completely negated. Market Driven Innovation projects have little to no implication on the existing DDOM.

Market Driven Innovation projects are separate and distinct from another type of project typically seen in business, the projects that are unique to a customer typically called engineer to order (ETO). A key characteristic of ETO environments is that a high part of the components will be drawn, set up in a unique bill of material and routing, and will never be used in entirety again after delivery to the customer. We will call these projects "Customer Driven" projects. These types of projects are often the primary offering of a company but could also be a joint development opportunity to capture a new customer or gain additional long-term business from an existing customer. An example might be an offshore drilling platform for a specific oil exploration company. An ETO company's DDOM is all about project management and will utilize control, time buffers, capacity buffers, and to a limited extent, stock buffers on any common materials. These projects reside in the operational and tactical relevant ranges for an organization.

Customer Driven projects share many characteristics with Market Driven Innovation projects but with one critical difference. The most important distinction is that an ETO project starts with an actual customer order! Market Driven projects might result in additional future customer orders (that is the intention), but there is no immediate customer pressure or commitment. The target of Customer Driven projects is to create a specific something to be delivered to a specific customer. Typically, delays may result in penalties imposed by the customer.

The flow of both types of projects is largely based on the flow of information and integration. Regardless of project type or detail, the objectives of any project are universal: deliver the project on time, on budget, and within specification. As in manufacturing and supply chain management, the attainment of these objectives will be reliant on the flow of relevant information, materials, and services.

Integrating project management into the DDAE model will be the subject of future works.

Summary

This chapter in and of itself is somewhat of a summary of this book. It connects the in-depth explanation from Chapters 4 to 6 with the specific design and elements of the DDAE model to the very beginning premise of the book: that organizations must learn how to adapt or face extinction. The DDAE model fundamentally starts at a different place than a conventional management approach, always emphasizing flow over unit cost. It is specifically designed to produce more relevant information and consequently more relevant materials and services around this fundamental principle of flow, something that most great business thinkers have embraced or clamored for throughout the last century.

In Chapter 8 we will turn our attention to the implementation of a DDAE model; it is not to be taken lightly. The hardest part of implementing the DDAE model is taking the first step, to make the fundamental shift from efficiency based on cost to efficiency based on flow.

CHAPTER 8

The DDAE Model Development Path

The DDAE model has a well-defined development path for companies to achieve increasing levels of success leveraging a Demand Driven transformation. This path has five distinct stages.

Stage 1: Operational Efficiency (Cost)

The development path starts where most companies find themselves today—locked in a constant struggle trying to drive operational efficiency by controlling or minimizing cost. It's not that the importance of flow goes unrecognized in these companies, but any flow-based metric such as on-time delivery constantly struggles against directly competitive cost-based metrics and objectives. As described in Chapters 1 and 2, this constant conflict or tension will distort relevant information, directly leading to even more variability that breaks down flow. This is a recipe for extinction in today's hyper-competitive and volatile markets. The characteristics of a company in Stage 1 are listed in Figure 8-1.

Stage 1: Operational Efficiency (Cost)				
Operational Objectives	**Demand Driven Characteristics**	**Primary Metrics**	**Analytics**	**Personnel Capabilities**
Cost reduction and responsiveness	Conventional MPS, MRP, DRP and MES practices. Demand Driven principles are limited to the incorporation of actual demand into supply order generation in the demand time fence. There is a strategic conflict between cost and service.	OEE, fully absorbed unit cost, service levels	Absorption rates, total days of inventory, OTD and/or fill rates	Traditional supply chain management and cost accounting approaches

FIGURE 8-1   Companies stuck in the cost perspective

Readers wishing to understand more about how devastating and inappropriate the emphasis on cost is should consider reading these books:

- *The Goal—A Process of Ongoing Improvement* (30th Anniversary Edition), Goldratt and Cox, North River Press, 2014
- *Demand Driven Performance—Using Smart Metrics*, Smith and Smith, McGraw-Hill, 2014

Appendix B provides an example of a company in Stage 1.

Stage 2: Operational Efficiency (Flow)

Stage 2 begins a company's transformation into a Demand Driven Adaptive Enterprise. Moving from Stage 1 to Stage 2 requires a dramatic philosophical shift in thinking and understanding about what is truly "efficient" from a system perspective. This shift is not trivial as it requires a fundamental break from the conventional emphasis on cost supported by the metrics and strategies promulgated by that emphasis. Stage 1 assumes that ROI improvement is connected to better cost performance (a fallacy), while Stage 2 makes the critical connection between flow improvement to better ROI. These two views are not compatible with each other—they are, in fact, in direct opposition to each other. This point was made in Chapter 2 with Figure 2-2. This is a significant strategic conflict that often requires a larger system analysis for company leadership to fully understand how devastating trying to live in two worlds really is. Appendix B details the tools and process for this type of systemic analysis and uses this Stage 1 problem as an example.

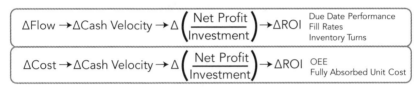

$$\Delta \text{Flow} \rightarrow \Delta \text{Cash Velocity} \rightarrow \Delta \left(\frac{\text{Net Profit}}{\text{Investment}} \right) \rightarrow \Delta \text{ROI} \quad \begin{array}{l} \text{Due Date Performance} \\ \text{Fill Rates} \\ \text{Inventory Turns} \end{array}$$

$$\Delta \text{Cost} \rightarrow \Delta \text{Cash Velocity} \rightarrow \Delta \left(\frac{\text{Net Profit}}{\text{Investment}} \right) \rightarrow \Delta \text{ROI} \quad \begin{array}{l} \text{OEE} \\ \text{Fully Absorbed Unit Cost} \end{array}$$

FIGURE 2-2   Flow versus cost perspectives and metrics

This shift into Stage 2 is initiated by the implementation of Demand Driven Material Requirements Planning. The question that many people ask is why this dramatic shift to flow starts with a different planning process rather than with some big strategic consulting project or technology investment. As evidenced in Chapter 1, the world has shifted dramatically as supply chains have elongated and grown more complex while customer tolerance times have shrunk dramatically. Strategically placed and well-managed stock positions are a necessity for every supply chain. As much as it may pain us to admit it, we live in a make to stock world at the shallow levels of most supply chains.

DDMRP represents the least amount of system "shock" and resistance in beginning to prove the beneficial difference of the Stage 2 flow emphasis over the Stage 1 cost emphasis. Planners and buyers, once skeptical of another new "improvement" method of the year, tend to quickly embrace DDMRP. DDMRP is a methodology that gives people a step-by-step blueprint that is transparent, easy to interpret, intuitive, consistent, and sustainable. DDMRP solves the problems these planners and buyers are dealing with daily and that they realize they cannot solve with the current approach. Let's explore these important DDMRP characteristics one at time.

Transparent—characterized by visibility or accessibility of information especially concerning business practices.[1]

People tend to mistrust what they don't understand or can't see—especially when their jobs are on the line. Planners and buyers with any amount of experience will go to extreme lengths to try to gain visibility to relevant information—they will not just blindly do what the computer system tells them to do. DDMRP was developed with this characteristic in mind. Its equations are simple, intuitive, and straightforward, allowing planners and buyers to immediately determine why planning and execution recommendations are or are not being suggested. DDMRP's planning equation answers, in an elegant, straightforward format, the four critical questions planners and buyers really care about:

- What do I have?
- What do I have in the pipeline coming to me?
- What demand do I need to fulfill immediately?
- What future demand is relevant?

Easy to Interpret—to give or provide the meaning of; explain; explicate; elucidate.[2]

DDMRP uses intuitive colors and percentages to indicate the current and future status of critical protection points. Color gives a general reference while the percentage gives a discrete reference for prioritization. This allows planners and buyers to quickly determine which items require resupply or attention. Figure 8-2 shows planning and execution priorities from DDMRP. Obviously, this text is printed in a monochromatic format that diminishes the powerful impact these views can have. In the electronic version of this text a reader has the benefit of seeing the colors. The top view is of planning priority; it indicates the net flow status and the severity of the need to launch a supply order for replenishment of the position. In this case, the part 406P is in critical need of additional supply orders. The bottom view is focused on the current on-hand position of a buffer relative to its safety zone (the red zone). It shows which positions need the most attention right now; typically, through an expedite of existing supply orders.

Today's Date: 15-July											
Part#	Planning Priority	On-Hand	On-Order	Qualified Demand	Net Flow Position	Order Recommendation	Request Date	Top RED	Top YELLOW	Top GREEN	Lead Time
406P	RED 19.8%	401	506	263	644	2606	4-Aug	750	2750	3250	20
403P	YELLOW 43.4%	1412	981	412	1981	2579	23-Jul	1200	3600	4560	8
402P	YELLOW 69.0%	601	753	112	1242	558	24-Jul	540	1440	1800	9
405P	YELLOW 74.0%	3400	4251	581	7070	2486	24-Jul	1756	7606	9556	9
401P	YELLOW 75.1%	2652	6233	712	8173	2715	25-Jul	2438	8938	10888	10
404P	GREEN 97.6%	1951	1560	291	3220	0		1050	2550	3300	6

Order #	On-Hand Buffer Status
PO 819-87	27% (RED)
WO 832-41	42% (RED)
WO 211-72	88% (YELLOW)

FIGURE 8-2   DDMRP planning and execution views

Intuitive—using or based on what one feels to be true even without conscious reasoning; instinctive.[3]

Because DDMRP is transparent and easy to interpret, most people find it highly intuitive. An inexperienced planner or buyer can have immediate effectiveness within a DDMRP system. An experienced planner or buyer can refocus their time and leverage their experience to work on improving the DDMRP model's overall performance and therefore directly contribute to driving better company ROI. This marks the beginning of DDS&OP activity.

Consistent—constantly adhering to the same principles, course, form, etc.[4]

The DDMRP approach allows planning and purchasing personnel to speak the same language across a department, multiple plants, and even companies. This eliminates the typical individualized or fragmented approaches to planning and execution that are commonplace in most companies. With this consistency comes better potential for consensus and clarity about new directions and innovations for the supply chain function. This consistency also allows planners to take a vacation without worrying about their parts.

Sustainable—able to be maintained or kept going, as an action or process.[5]

A key for sustaining anything is that people must continue to use it, and people use what they trust. People will trust something more when it is transparent, easy to interpret, intuitive, and consistent. Furthermore, these attributes make it much easier to sustain performance with other personnel when critical personnel leave or are temporarily unavailable. This just makes sense from a risk mitigation perspective. This also has the impact of dramatically reducing stress for planning personnel, and with it, reducing personnel turnover.

Figure 8-3 depicts the attributes to Stage 2. The primary metrics and analytics are specific and limited to DDMRP and fit under the operational metric objectives described in Chapter 4.

Stage 2: Operational Efficiency (Flow)				
Operational Objectives	Demand Driven Characteristics	Primary Metrics	Analytics	Personnel Capabilities
Flow protection and promotion	Trial and/or expanding implementation of Demand Driven Material Requirements Planning (DDMRP)	Signal integrity, decoupling point integrity, average inventory, service	OTOG % and $, % to inventory target, OTD and/ or fill rates	Personnel are aware of and capable of describing the problems with conventional planning systems. They are well versed in DDMRP principles and are capable of implementing (at a cursory level) decoupling point buffers

FIGURE 8-3   Stage 2 attributes

It is important to note that DDMRP is only the beginning of a DDAE transformation, not an end. It is a significant starting step, a beach-head, to a much larger transformation into becoming a Demand Driven Adaptive Enterprise. DDMRP quickly produces significant financial and performance results, reinforcing and proving the doctrine of flow. DDMRP provides the confidence for an organization to keep going—to continue to strip out the common nonsense and build practical management systems and metrics based on flow that facilitate true return on investment in a highly complex and volatile world.

Can initial DDMRP efforts fail? Of course. As in any new initiative, the organization may not be ready to fully bring a DDMRP effort to fruition. In the authors' experience, these efforts have failed for the following reasons:

- The premature loss of a project champion. In the past (the early years of Demand Driven), the initial shift to Stage 2 typically occurred at a relatively local level (plant) and was led by a local champion implementing DDMRP principles in a limited fashion and even somewhat under the corporate radar. The loss of this champion would typically cause the effort to stall even with substantial positive results from the pilot implementation. As Demand Driven has recently become more mainstream, however, the transition to Stage 2 is now being driven from the top, typically in the form of sanctioned pilots at the plant and/or product line level with dedicated and mandated resources supported by widespread conceptual education.
- The lack of support from another critical and interdependent function. In larger organizations with bigger functional silos, it is important to understand what can affect the success of an initial DDMRP effort. An example: in 2013 one of the authors was working with a large consumer products company helping to implement DDMRP concepts across several factories in North America. The factories had implemented DDMRP principles on purchased materials and were beginning to implement on the finished prod-

uct side. There was a need to move to a hub and spoke configuration for distribution in the United States. The initial results
of a hub and spoke configuration showed tremendous inventory
leverage with significant financial and performance improvement
in Canada; the United States promised a much larger leverage. In
the middle of this transition two things occurred. The first was a
large acquisition of a well-known brand and all the inflated inventory that came with it. This dramatically reduced the warehouse
space available for hub stock positions at a critical location. The
implementation team was able to mitigate that circumstance with
some creative innovation. The next move, however, was a knock-
out punch. The Deployment division, in a move to reduce transportation cost per unit (their primary metric), began to realign
and redesign the distribution network, throwing the DDMRP
design and manufacturing schedules into chaos. It took nearly one
year to get Deployment to understand what was happening and
how much damage had been wrought in the organization due to
their cost fixation.

- A reinforced emphasis on cost. Some organizations have an event
or periodic public measurement that may cause panic from the
executive level, resulting in an edict to immediately emphasize
cost performance. This can and often will cause significant disruption to any DDMRP or any flow-based efforts and significantly
jeopardizes the company ROI in the process.

- The lack of an integrated IT solution. In the early years of DDMRP,
software was not readily available. When available, it was typically difficult to get IT on board with interfacing "another add-on
application" to the ERP product. IT would fight this initiative with
all its might. As DDMRP became more popular, software companies (including many large and small ERP and best of breed
add-on application provider software companies) began to embed
DDMRP principles into their products. In 2014 the Demand
Driven Institute launched a DDMRP software compliance program to ensure that software meets a minimum specification to

adopt and sustain DDMRP. That DDMRP software compliance criteria is described later in this chapter.

When the initial DDMRP efforts fail, any DDAE transition is essentially halted; you simply cannot have a Demand Driven Operating Model without DDMRP. Most organizations will simply slide back into Stage 1.

DDMRP is not a silver bullet. It is, however, a more appropriate, safe, and reliable planning and execution method that reduces inventory, improves customer service, and reduces planner stress. This provides the organization with the confidence to proceed further by expanding the DDMRP implementation to the enterprise level and eventually moving to the next stage of the DDAE Development Path.

Stage 3: DDAE Level I

Stage 3 is the first level in which an organization can really begin to describe itself as "Demand Driven"; thus the name of the stage is DDAE Level I. DDAE Level I consists of a fully implemented Demand Driven Operating Model (DDMRP, Demand Driven Capacity Scheduling and Demand Driven Execution methods in use). It should be noted that in some environments, DDMRP alone is sufficient to constitute a fully implemented DDOM. These types of environments often include batch processors with simple routings and/or limited diversity of resources such as injection molders, bottlers, or any operation with single stage processing, manufacturing, or assembly. In these environments, the maturation of the DDMRP implementation would be synonymous with the maturation of the DDOM.

Yet the majority of organizations require more than DDMRP to have a fully implemented DDOM. The movement from Stage 2 to Stage 3 (DDAE Level I) can take several years in larger organizations with multiple facilities, complex routings, diverse sets of resources, and vertical integration. This represents an extensive (but significantly beneficial) overhaul of operating tactics impacting supply order generation, resource scheduling, operational execution, and metrics. This stage is thoroughly described in *Demand Driven Performance—Using Smart Metrics* (Smith and Smith, McGraw-Hill, 2014). A maturing Stage 3 company will typically begin to become constrained by a lack of alignment from other functions in the organization to flow and the DDOM design.

Figure 8-4 shows the characteristics of a Stage 3 company. The figure shows the emphasis placed on the operational relevant range metrics described in Chapter 4 as well as the tactical analytics (past tactical relevant range) required to begin DDS&OP activities.

Stage 3: DDAE Level I				
Operational Objectives	Demand Driven Characteristics	Primary Metrics	Analytics	Personnel Capabilities
Fully synchronize and leverage operational capability for better flow performance	Trial and/or expanding implementation of the Demand Driven Operating Model (DDOM) with supporting tactical analytics	Reliability, stability, velocity	Buffer run charts, reason code analysis, flow exception reports, flow indices	Personnel understand the broader implications of DDMRP to the organization. Personnel understand how to implement Demand Driven Scheduling and Execution. Personnel are capable of adjusting the DDOM based on performance analytics.

FIGURE 8-4   Stage 3 characteristics

Stage 4: DDAE Level II

Stage 4 (DDAE Level II) describes the expansion of the Demand Driven concepts throughout the organization. A tactical reconciliation process is in place with DDS&OP and the entire organization understands how to leverage their mature DDOM capability into the market and throughout the organization for better financial performance. Its personnel understand and see the company as a holistic system. Finance, Engineering, IT, Marketing, Sales, and Strategic Planning understand how to use the DDOM as a competitive advantage and can communicate through a common flow-based language; typically, led by a cross-functional DDS&OP team.

Stage 4: DDAE Level II				
Operational Objectives	Demand Driven Characteristics	Primary Metrics	Analytics	Personnel Capabilities
Leverage the Demand Driven Operating Model capability across the enterprise and into the market	A mature DDOM with the strategic and tactical reconciliation process of DDS&OP with Adaptive S&OP in place. A full flow-based metric suite in place.	Strategic contribution, waste/ improvement, operating expense control, RACE/ ROIC/ROI	Outlier analysis (time, capacity and stock buffers), buffer compression, contribution margin rate and volume improvement	Other functional personnel now understand the requirements and capabilities of the DDOM. Personnel can successfully bridge the tactical and strategic relevant ranges. They can project, recommend, and adapt.

FIGURE 8-5   Stage 4 attributes

Figure 8-5 displays the attributes of a Stage 4 company. The figure shows the metrics, including the tactical metrics described in Chapter 5, in addition to some form of return on investment being watched from higher levels. The tactical and strategic adaptive loops are in place and the organization is learning how to use them and bidirectionally reconcile them effectively. This drives internal innovation, reveals opportunities, and allows for effective scenario flow-based analysis. Chapters 5 and 6 describe these kinds of activities. The organization is now sensing and adapting and is ready to go to the next level.

Stage 5: DDAE Level III

Stage 5 (DDAE Level III) describes how the organization can become a valuable and strategic supply chain partner by facilitating flow between itself and its suppliers and customers in mutually beneficial ways. Its personnel understand and see the supply chain as a complete interconnected network identifying opportunities for better flow creation and protection. Management has the capability to define current and/or impending strategic conflicts and reconcile them through adaptive and innovative solutions. These organizations are capable of mentoring new generations of management through the DDAE model in order to sustain and even accelerate momentum.

Figure 8-6 describes the characteristics of a Stage 5 enterprise with the ultimate objective of driving better rates of shareholding equity (both now and in the future). The analytics are an interesting note. How does an organization know that it continues to drive to its strategic long-term mission? This can often be tracked by its ability to identify and subsequently resolve large strategic conflicts. These conflicts create waste or "friction" that ultimately reduces flow or blocks a significant opportunity for the enterprise. These types of conflicts can be internal or external to the enterprise.

Stage 5: DDAE Level III				
Operational Objectives	**Demand Driven Characteristics**	**Primary Metrics**	**Analytics**	**Personnel Capabilities**
Sense, adapt, and innovate across the organization and supply chain (customers and suppliers)	A mature DDOM with mature DDS&OP and Adaptive S&OP. Thoughtware fully installed.	RACE/ROIC/ROI improvement rate	Strategic conflict definition and resolution	Strategic personnel can analyze complex problem areas (internal and external), define strategic conflicts and constraints and recommend strategic policy/ direction changes. They are able to mentor new key personnel through the DDAE Model.

FIGURE 8-6   Stage 5 attributes

To identify and break through these strategic conflicts, adaptive agents must be prepared to think and resolve issues from a systemic perspective. This process can and should be learned by most strategic adaptive agents. Appendix B describes this process in more depth.

Software Implications

The DDAE model has major implications for software. The Demand Driven Institute has prepared a set of specifications for software to facilitate and support the transformation to a Demand Driven Adaptive Enterprise. Conventional planning tools will not get a company to even Stage 2. For an in-depth explanation why, the reader should consider reading *Precisely Wrong—Why Conventional Planning Fails and How to Fix It* (Ptak and Smith, Industrial Press, 2017).

Demand Driven Operating Model Compliance Criteria

The following criteria has been established to support Stage 2 and 3 of the DDAE development path, beginning with DDMRP compliance.

Demand Driven MRP Compliance Criteria

The Demand Driven Institute established the following basic criteria for software to be compliant to the DDMRP method. The compliance criteria are intended to ensure that a piece of software has enough features and/or functions to implement, sustain, and even improve a DDMRP implementation. The criteria were written in such a way that it ensures compliance to the fundamental principles of the method but allows enough open space for competitive differences, creativity, and innovation for the software company.

Component 1: Inventory Positioning
- The software must be able to calculate and identify the decoupled lead time for manufactured items.
- If the decoupled lead time calculation cannot be performed, then DDMRP compliance will be limited to Purchased and Distributed Parts only.

Component 2: Buffer Profiles

- The software must be able to group parts into independently managed families with variable settings for zone impact.
- The software must be able to calculate DDMRP buffers and zone values using a combination of buffer profile attributes and the individual part traits of usage, lead time and order multiple and/or order cycle.

Component 3: Dynamic Buffer Adjustments

- The software must have a provision for dynamically altering buffers for planned or anticipated events.

Component 4: Demand Driven Planning

- The software must be able to perform the DDMRP net flow equation including qualifying sales order demand properly (due today, past due, and qualified spikes).
- The software must be able to properly display net flow status (color, percentage, and quantity) for easy prioritization and supply order generation.
- All elements of the net flow equation should be visible and/or easily accessible on the planner workbench.

Component 5: Highly Visible and Collaborative Execution

- The software should display alerts based on the on-hand buffer status for decoupled positions.

Of course, more complex, larger entities will need deeper and richer features. This list represents only the basic features that any DDMRP compliant planning and execution system should have.

The Demand Driven Institute is responsible for evaluating packages against these criteria. Since the introduction of these criteria in 2014, a number of software packages including popular ERP products have become compliant. This would not be happening unless the market was beginning to demand something different than conventional planning; in particular DDMRP functionality.

Demand Driven Scheduling Compliance Criteria

In addition to DDMRP functionality, software supporting a full DDOM approach will need to comply with additional minimum scheduling and buffer management criteria.

Component 1: Additional DDOM Modeling

- The software must be able to declare identified resources, control points, or drums.
- The software must be able to place time buffers at defined points in routings and/or in advance of resource activity.

Component 2: Demand Driven Scheduling

- The software must offer finite capacity scheduling for control point resources or sets of resources.
- The software must be able to tie the finite scheduling of control point resources to order due dates and/or net flow priorities of stock buffers.
- The software must synchronize material release dates, times, and quantities in relation to the control point schedules including allowances for time buffers.

Component 3: Demand Driven Execution Requirements

- The software must track order progression against time buffer penetration and control point schedules.
- The software must allow for reason code collection at buffers for outlier events.
- The software must display detailed capacity loading for all resources.
- The software must display or connect the stock buffer status that an order is progressing toward.

Demand Driven Sales & Operations Planning Compliance Criteria

The Demand Driven Institute has developed this list of minimum compliance criteria for DDS&OP. Examples of these criteria were described and depicted in Chapter 5.

Component 1: Master Settings Access

- The software must provide access to all relevant master settings of the DDOM.

Component 2: DDOM Variance Analysis Capability

- The software must be able to produce reports on past buffer performance for stock, time, and capacity buffers including statistical run charts

Component 3: DDOM Simulation Capability

- The software must be able to simulate DDOM performance regarding capacity and working capital levels under different user defined scenarios.

An up-to-date list of DDMRP, DDOM, and DDS&OP compliant software can be found at: www.demanddriveninstitute.com/compliant -software.

Summary

What stands in the way of Demand Driven global proliferation is a series of common conventional practices and assumptions in both Operations and Finance that must be understood for what they really are—common nonsense. Optimizing these old and inappropriate rules in a more complex and volatile set of circumstances will only push organizations further away from embracing flow while incurring devastating amounts of waste, resulting in eventual company failure.

The Demand Driven Adaptive Enterprise Model is first and foremost about visibility to what is relevant. It recognizes that the only way to effectively implement and foster flow is to enable a company to determine truly relevant information at the strategic, tactical, and operational levels. Through that visibility, companies can also eliminate what is irrelevant, distortive, and damaging.

It is the authors' firm belief and experience that the key to start and sustain this DDAE transformation is rooted firmly in education on the principles and methodology. Companies should be careful about emphasizing speed of implementation over foundation of implementation. This foundation is built on the right education for the right personnel and ensuring the appropriate implementation pace for the unique circumstances of the organization. There are two points at opposite ends of the spectrum that all airplane pilots fear: stalling and disintegration. If a plane travels too slowly there is not enough lift generated and it falls from the sky. If a plane travels too fast, friction literally tears it apart and it falls from the sky. Flying is about understanding and maintaining the appropriate pace given the unique circumstances and characteristics of the plane and the mission/objective. Driving large transformations in organizations is no different.

A complete journey through these five stages can take several years. The upper stages (4 and 5) may never be achieved as key personnel exit for other opportunities (especially the project sponsor) and/or acquisitions occur that slow the momentum of the DDAE model implementation. Not taking that first step, however, guarantees that the organization will be stuck and struggle to appropriately adapt. At each step or stage of the DDAE model ROI improves and accelerates. So, what does your organization have to lose? Management truly has two choices: adapt or die.

Skill Buffers: The Missing Links

by Caroline Mondon

Much has been written about how to manage stock, time, and resource capacity in the Demand Driven methodology, but what about the people and skills that it takes to maintain and sustain a Demand Driven Adaptive Enterprise?

All companies, whatever their size and type of industry, experience symptoms of flow disruption due to missing skills. The disruption can be more or less significant depending on how often the skills are missing and how long it takes to find an alternative. Disruption can happen because the only skilled person is sick or on leave. It can happen because management didn't anticipate resignation or retirement and didn't organize the transference of the skill. Alternatively, this can happen because sharing a skill is perceived as a loss of power by the experienced expert who thinks that being the only one to master that skill ensures job security. Even if that situation may cause a lot of stress and burnout for the owner of this particular skill, the individual may prefer to keep a stranglehold on a particular skill because it makes them indispensable. This is especially true when the company doesn't have a formalized system to recognize and value the long-time experienced employees who share their skills.

In most companies, the only path to grow in salary and recognition is to become a manager or a leader. Traditionally, it is the person that has a higher level of skill and experience that advances through the ranks. With this rise comes better compensation and benefits. Unfortunately, this path leads to the well-known effect called the "Peter Principle"[1] where the promotion process results in gaining a bad manager and losing a good technical expert. Many employees help or coach their colleagues because they are naturally generous. But after many years of doing so, if they are not recognized for their contribution, the improvisational process to educate other people results in a general feeling of frustration and disappointment. In such companies, if these employees are ambitious and want to get a raise in salary or recognition, the best tactic is obviously to retain the unique and valuable skill so that management has no choice but to give a raise in salary with the title of manager that goes with it.

Small or medium size companies, where there are few management or leadership positions, often have to deal with a layer of middle management who got their promotions thanks to having a general reputation of being "super man or woman." Super men or women are managers or leaders who continuously feel the need to prove that their involvement in everything is vital for company success. This often leads them to extinguish fires that have been self-ignited. For example, they expedite orders because the frozen schedule is not valid compared to actual demand and they cannot keep up with updating it. Another example is trying to solve quality problems due to new technical specifications where the validation process is late because it needs their signature to proceed, or because the new employee doesn't know what to do and is waiting for their instructions, etc. In such a context, if the company wants to become demand driven, its middle management is often a major bottleneck. This resistance is not because the managers are overloaded but because the management system doesn't motivate them to share their skills. These middle managers or leaders are usually the most attached to the company because their ego is based on their super man or woman image in that particular company. They were seldom given the time to formally acquire new skills because they are constantly overloaded, and they didn't take the time to learn new skills and become certified to demonstrate value in the market the day they want to find another job. They are indeed stuck in a vicious cycle.

Fortunately, in more and more companies, continuous education is transforming companies into learning enterprises. In order to adapt to new technologies and market constraints, education is valued and must be systematically applied. Many governments financially sponsor annual plans of continuous education. However, when the employees only know their segment of the overall flow where their skill is relevant, they don't really know what the market is expecting because it changes more and more often in a VUCA (Volatile, Uncertain, Complex, Ambiguous) world. Employees cannot guess what order winner is valued by the customers when they have never met any of them. How can they understand the systemic paradigm of the Demand Driven Enterprise Model? How can they contribute to protect and promote flow with the appropriate intuitive actions? In such a foggy context, the best that management can hope for is to motivate their employees to learn how to improve what they are already doing to contribute to department goals. They indeed get trapped in silos.

The intention of this appendix is to describe the missing links for a systemic continuous learning path for all employees of the company that will enable them to address:

- Bottleneck of skill in the organization
- Guidelines to protect and promote flow at individual level
- Vision of the strategic skills priorities to share
- Metrics related to skills sharing
- Multiple paths to grow in salary and recognition
- Win-win relationship between employees and companies for continuous learning
- Valuable transition for the most experienced employees and for the redundant middle management

Managing skills with a systemic approach facilitates change toward a Demand Driven Adaptive Enterprise when progress is visible, relevant, and valuable for the organization's most important assets—its human capital.

Defining a Skill Buffer

As this book has described in detail, the first step in a demand driven transformation is the positioning of buffers. These time, stock, and inventory buffers, as shown in Figure A-1, absorb variability from both sides and enable the enterprise to sense changes in demand, adapt planning and production, and pull from suppliers in real time. But what about the skills required at the different resources? In order to protect against variability due to skills availability, consider a new buffer that is to human beings what a capacity buffer is to machine resources. A capacity buffer protects flow by providing available capacity to catch up with variability. It is a protective buffer that provides agility and flexibility. The size of the capacity buffer allows both stock and time buffers to be reduced. It requires that a resource maintain a bank of capacity that may often go unused.

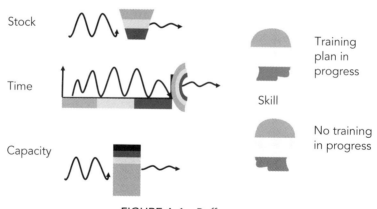

FIGURE A-1   Buffer types

The similarity between machines and human beings stops here. Machines can keep their complete range of capability for a long time, even if not used. Human beings need to exercise their skills frequently to be able to use them efficiently when needed. In Figure A-1, the symbol of the skill buffer (shaped as a human head with the standard green, yellow, and red) has two positions. When the head is turned toward the right, it means that a training plan is in progress to maintain continuous development of that particular skill. When the head is turned toward the left, no training is scheduled for that skill.

When an internal trainer is qualified to teach a skill to a student, that also provides the opportunity for this trainer to validate new specifications or process descriptions. The trainer-student team is usually the one who has the relevant practical information to validate a process when their manager or team leader has only a theoretical understanding of the constraints. This allows managers to focus on the tactical performance rather than the detailed operational performance.

A difference between the standard buffers and this new skill buffer is the meaning of the three colors. The priority order remains the same.

- **Green.** All is good. The quantity and level of skills available are sufficient to face the variability of demand for this particular skill. There are enough employees who can apply the skill; an expert who knows all about this skill, and a trainer who is in the process of teaching a student or coaching an operationally qualified individual toward expertise on the skill. An employee is operationally qualified when he or she is not fully autonomous but can proceed with at least 20 percent of the total skill that supports 80 percent of the activity at the workstation.
- **Yellow.** There is a risk identified because while there is more than one person able to apply the skill, there is insufficient contingency for variability. The legend for this color as well as for the other colors can be adapted depending on the level of maturity of the company. For example, this level could mean at least one expert who knows all about the skill or a trainer is available. For some skills, operational but not exhaustive knowledge of the skill may be acceptable. A yellow level skill buffer could also mean that an insufficient number of trained resources are available to support

innovation in the specifications of the product and service or in the process.

* **Red.** The resource capacity is at risk. For example, if there is only one person who knows 100 percent (or even less) of the skill, flow is at risk in the short term. If the skill is only mastered by experts who are not competent to train others, the risk is not immediate but still of concern in the medium or long term. The company may not be able to innovate or adapt to new specifications of the product, the service, or the process.

Each of these levels is adapted depending on the situation of the company. Figure A-2 shows how the skill buffer can be symbolized by a colored head. The red head on the left illustrates a critical situation of missing skill, no trainer available and no training plan in progress. If the company decides to train the employee in a public class, the red head will turn toward the right. The green head illustrates a situation where there is a trainer available. If no training is in progress nor scheduled, the green head will be turned toward the left to bring awareness. The symbol of the head is typically used with plain colors.

Red Yellow Green Head
No Training in Progress

Red Yellow Green Head
Training in Progress

FIGURE A-2   Head symbol of skill buffers

Segmenting skills of the company into functions highly depends on the size of the company and on how the processes are distributed among the employees. Companies who start the process of buffering skills will need to define many detailed skills and functions, often in silos, even in the same department. The more the company focuses on protecting and promoting flow, the more detailed skills will be grouped as generic skills as the employees become more and more multiskilled. This evolution simplifies capacity management and therefore scheduling. This enables the company to quickly adapt to changes in demand.

As an example, at the beginning of the demand driven detective novel *The Missing Links*, two onsite installation workers are specialized in driving the delivery truck for luxury furniture for boats and fancy clothing stores and in installing this furniture on site. At the end of the novel, being able to install the furniture on site is part of the training plan to become "operational woodworker or metalworker." This was done to provide the opportunity to better understand customer constraints and requirements. Therefore, onsite installation and woodwork or metalworker is considered as one skill.

The Demand Driven Skill Model

Now that there is visibility to specific skill buffers, let's examine how the multiskill grid can be transformed into a dynamic Demand Driven Skill Model (DDSM). The DDSM enables a company to visualize in real time what the training priorities are to protect flow.

The multiskill grid is well understood by companies certified ISO9000 because it is one of the requirements. The DDSM builds on this tool to describe the multiskill of all employees, including top management, in order to focus on flow promotion and protection. The DDSM encourages transmission of knowledge, continuous improvement of products and processes, and integration of innovations by a team of internal trainers who support and complement the management team.

Building the Multiskill Grid

How do you describe the level of employee capability to protect and promote flow in your company? Figure A-3 is an example from a manufacturing company of the capability continuum from student to internal trainer. It must be adapted to each company depending on:

- Maturity regarding flow focus/cost focus

- Education on best practices (Lean, 6 Sigma, TPM, TQM, etc.) and understanding of how each can contribute to protecting flow
- Intimacy with customers
- Complexity of the machines and of the technologies used
- Empowerment given to employees and therefore role of the managers and leaders: are they in a supervisor or coach posture?
- Silo thickness between departments such as: finance, supply chain, total quality

Each level builds on the skills of the previous level so the incremental skills are shown on the diagonal.

| Student | Operational | Expert | Trainer |

Understand Flow	Promote Flow	Protect Flow	Design Flow
			Design and coach 100% of the system
		Manage 100% of the job	
	Manage 80% of the job		
Understand system components			

FIGURE A-3 Multiskill grid definition of levels

A student is learning about flow and understands the system components:

- Markets, customers, product lines, shareholders, financial situation of the company
- Specifications of the company products and services
- Health and security rules of the company
- Basic lean methods such as 5S and TPM-like first level maintenance
- Basic quality control tools

An Operational employee can manage 20 percent of the total skill that supports 80 percent of the activity at the workstation and promote flow with the appropriate good practices like:

- Total Quality Management
- Buffer board status
- Control point schedule

An expert can manage 100 percent of the job and is able to protect flow and contribute to:

- Problem solving to ensure quality of all products and services
- Management of all types of appropriate buffers

An internal trainer can teach or coach 100 percent of skills and is able to:

- Consider the systemic view to innovate successfully regarding products and services
- Create an educational path from beginner to expert
- Audit the process and update it when necessary
- Calculate the ROI of continuous improvement projects

A good practice is to first ask the employees to self-assess their own level before constructing the official multiskill grid. This enables the identification and resolution of possible significant gaps between employee perception and management's perception of the reality. An example is shown in Figure A-4.

The multiskill grid must have the attention of top management to encourage continuous learning and adaptation of skills to the strategy of the company and to the demand of the customers.

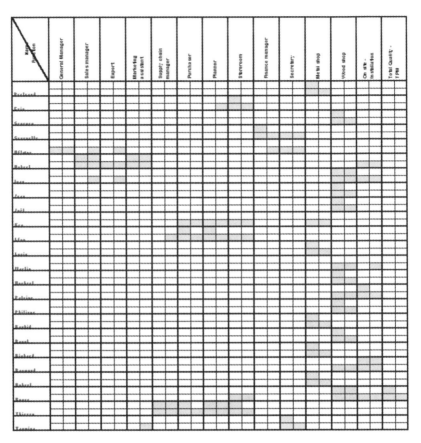

FIGURE A-4   Example of a multiskill grid with function
on top and names on the left

Valuing Skill Sharing

As shown in Figure A-5, dedicating a several meter square space in the plant for the multiskill grid provides a central location to celebrate new green squares. This is a visible demonstration that the topic is important for the future of the company.

Celebration of new trainers can become as important as celebration of new managers or leaders. When top management provides focus and leads by example, it takes at least one year to change the culture of a company toward valuing skill sharing.

FIGURE A-5 Example multiskill grid celebration
Drawing copyright Caroline Mondon. Used with permission.

To become an internal trainer, a specific public or in-house training of at least three days is required to teach an individual the specific skills necessary to be a trainer. Therefore, at the beginning of the project, no one can be a self-assessed trainer since this training has not been taken. This requirement provides a good way to discourage a well-known but ineffective internal improvised trainer; for example, people who believe that to explain better you should repeat more loudly.

Typically, a company will add customized days to this training to take into account the specific processes of the company. During these days, the instructor will coach the trainers-to-be to make a presentation of the detailed content of the training he or she designed for their peers. Union representatives are usually welcome to participate in this process in addition to top management.

During the face to face annual management review with the employees, managers, leaders, or HR must be trained to properly explain the consequences of the missing skill on flow. Priority skills are identified and made visible to inspire career paths. Figure A-6 is an example of a visibility chart that can be used during the face to face annual review process. Note in the figure that not only are the skills insufficient, which means the head is red, but also the head is turned to the left which communicates that there is not a training plan in place to address this deficiency.

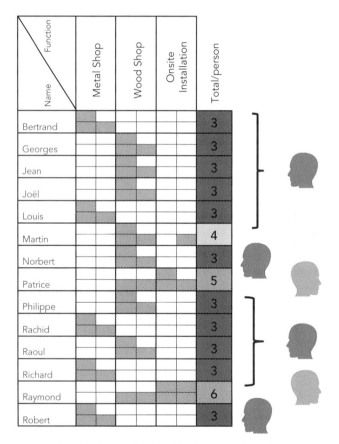

FIGURE A-6   Multiskill grid for individuals

Valuing Multiple Paths to Growth

One of the advantages of using the number of green squares as a new way to value the contribution of an employee is the variety of growth paths it offers employees that can be equally valued. Depending on individual aspirations, several paths can be proposed to new employees or to current employees that will motivate them to stay in the company and grow their talents.

In Figure A-7, all six configurations result in seven green squares for the individual. For example, if the internal trainer is very valuable to a continuous improvement project, a young employee who has some functionality across four functions can rescue a customer order when demand variability is higher than expected.

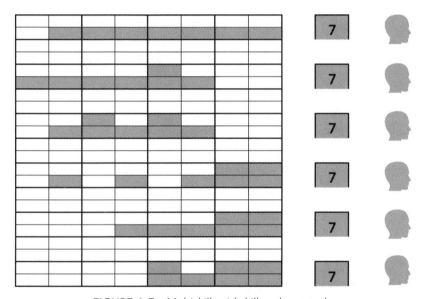

FIGURE A-7   Multiskill grid skill path example

Links Between Competencies and Competitivity

The ROI of training can be difficult to evaluate. It's one of the oldest jokes in the business world:

> Two managers are talking about training their employees. The first asks, "Yeah, but what if we train them, and they just leave?" The second responds, "What if we don't train them, and they stay?"

All companies looking for better financial results face the dilemma described in Figure A-8.

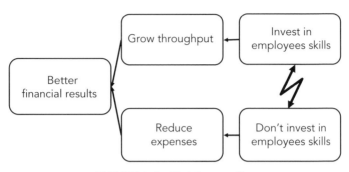

FIGURE A-8   Training conflict

They must grow their throughput and at the same time reduce expenses. Growing throughput requires well-trained employees and a continuous update of their skills. Reducing expenses cuts into training budgets. As a consequence, companies oscillate between training and not training depending on the manager in charge and other circumstances that either emphasize the pressure to reduce expenses versus adding skills and expanding skill levels. When training is postponed and finally occurs late in the year because of financial reasons, it will most likely have little linkage to promoting and protecting flow.

Enterprises already engaged in the Demand Driven Adaptive journey understand that a few days of internal or public training will never cost more than the consequences of a break in flow that can cause the loss of a customer.

Therefore, the priority will always be to focus on flow protection and to turn red heads toward the right; that is to say, invest in training a sufficient number of professionals to master 100 percent of those skills. Eventually some employees will be motivated to become internal trainers for that skill.

But once red heads have turned toward the right and become yellow, how then does a company decide on priorities to schedule the trainings?

The UIMM federation (Union des industries et métiers de la métallurgie), the largest French union of employers, put together a very simple way to formalize priorities in terms of training. A CEO who wants to get subsidies for training employees must answer on one page the questions in Figure A-9 called the Competency–Competitivity Plan.

Competitivity Target		New investment	New Competencies		Training Paths	
Actual	Future		New jobs	Evolution of internal skills	Year n1 to n3	External or internal	Internal trainer

FIGURE A-9   Competency–Competitivity Plan

To avoid the traditional silo effect of thinking about competitivity targets per department, five main transverse processes can be considered instead. Using transverse processes also provides more employee opportunities to become multiskilled between departments and to start new careers in addition to teaching the company to think systemically.

The first two transverse processes are Marketing and Sales, and Supply Chain. The common target of these two operational processes is to protect flow in the current environment to ensure proper delivery of orders to customers. The next two processes are Finance and Human Resources. The common target of these two support processes is to promote flow by anticipating the financial and human resources necessary to support the business plan with the relevant investment and skills. The last of the five transverse processes is Total Quality Management. The main responsibility of this transverse process is to formalize all processes of the company in order to guarantee customer satisfaction in safe conditions for the employees. Processes are validated with the use of audits and tactical metrics. The Total Quality function must also be able to adapt processes when necessary to deploy innovation. This is where the team of internal trainers becomes a strategic resource for the company to be able to innovate quickly and safely.

Figure A-10 is an example of what the company H.RAMI, symbolic "heroine" of the demand driven detective novel *The Missing Links*, planned for the next three years in the transverse process "Marketing-Sales."

Transverse Process	Competitivity Target		New Investment	New Competencies		Training Paths		
	Actual	Future		New jobs	Evolution of internal skills	Year n1 to n3	External or internal	Internal trainer
Marketing-Sales	Less than 5% of sales is export with only 1 international customer	New export markets in product lines Croisière and Boutique **Target:** 33% of export sales with at least 3 customers in 3 years in each product line			Multiskilled evolution of a woodworker (Ivan) into a export salesperson	Year n1 to n3: Business English	external by phone	
						Year n1: Apprenticing for Ivan	internal	Hubert
						Year n2: Export sales manager course	external	
	Our understanding of market needs, potential customers expectations, and competitors is not sufficient	Clarify order qualifiers and order winners of our products and services **Target:** Yearly focus customer and prospect group to update order qualifier and winner	CRM software after 1 year of manual monitoring	Bilingual secretary (to back up Yasmina)	Coaching of new secretary	Year n1: Apprenticing for new secretary	internal	Yasmina
						Year n2: CRM software training for Yasmina and new secretary	external	

FIGURE A-10 H.RAMI Competency–Competitivity Plan for the transverse process marketing-sales

The Competency–Competitivity Plan is most effective if it is succinct. If an executive committee cannot summarize on one page the target of the five transverse processes for expected priorities in terms of education, it is unlikely that the employees, the union, the managers, and the HR department will be able to deploy it.

How Can We Measure Progress of the DDSM?

If You Can't Measure It, You *Can't Improve* It.

—Peter Drucker

DDSM progress must be visual and intuitive to be attractive and to motivate all employees to maintain it. But if two head symbols are the same color, is it possible to see as easily as it is in the other DDI buffers what the priority is? Measurement of progress is made by calculating the average number of green squares by person and per skill after the head symbol has been colored to make priorities visual.

In Figure A-11 the company is looking for the missing skills in the Supply Chain and Total Quality–TPM departments, after assessing the situation per person and per skill in each department to prepare face to face annual management reviews with employees. The calculation of the average number of green squares/skill for the department is obtained by dividing total number of green squares per department/number of skills identified in the department. The highest priority is obvious; flow is at risk because the Total Quality–TPM department has only one skilled employee. If Roger is sick, no one can address a rework that must be expedited for the customer. In addition, there is no internal trainer (four green squares) in this Total Quality–TPM function. This means that no improvement of the processes of this department is possible as there is no training opportunity. As a result, the priority is to convince Roger to take a trainer course and then to find an employee interested in learning about Total Quality and TPM.

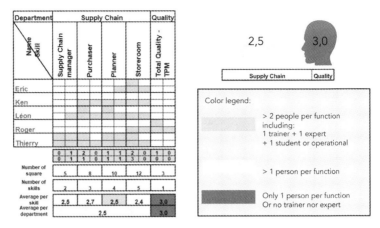

FIGURE A-11 Example skill buffer chart analysis—Step 1: focus on flow

Figure A-11 demonstrates Step 1: Focus on flow. It is clear that in the purchasing function, flow is not at risk because two employees, Léon and Thierry, both experts with three green squares, can manage 100 percent of the job. One employee, Ken, is operational for 80 percent of the job (two green squares) but there is no internal trainer mastering this skill. Therefore, this operational employee is not going to evolve quickly and there is no continuous improvement happening, nor future opportunities to implement innovations while teaching this skill.

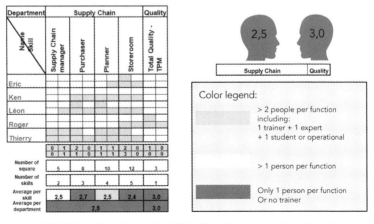

FIGURE A-12 Example skill buffer chart analysis—Step 2: focus on flow and innovation

Figure A-12 demonstrates Step 2: Focus on flow and innovation. Now the company wants to focus on flow and also on innovation. To reach that goal, the red legend has evolved from asking for a trainer or expert to asking for a trainer only. Therefore, both the Supply Chain and the Total Quality–TPM departments get red heads that both drive attention when deciding on priorities for the training plan. Imagine that Léon becomes a trainer in purchasing and coaches Ken to become an expert. The purchasing function can then be empowered by management to take initiatives. The purchasers' manager, Thierry, will save time that he can invest in becoming a trainer in the Storeroom. If Roger becomes trainer in Total Quality–TPM and trains Eric, the updated metrics are shown in Figure A-13.

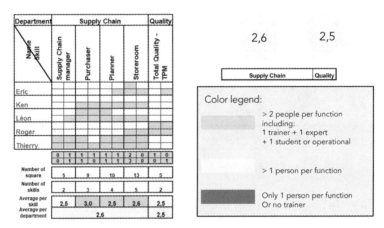

FIGURE A-13   Example skill buffer chart analysis—Step 3: focus on flow and innovation

The priority goes to the lower yellow number, the Total Quality–TPM department. If Roger is away, the student Eric will only be able to alert management about problems. In the Supply Chain department, if Thierry is away, the team now has enough trainers and experts to be able to react to variability without their manager's approval because they are skilled and autonomous.

Using DDS&OP to Link Adaptive S&OP and DDSM

Strategic Positioning of Skills

Once all employees are comfortable (and proud) with their respective green squares in the multiskill grid and when the education plan to achieve the competitivity targets is clearly described in the Competitivity–Competency Plan, then the skill buffers model can be displayed as shown in Figure A-14 to communicate the overall operational schematic with respect to strategy.

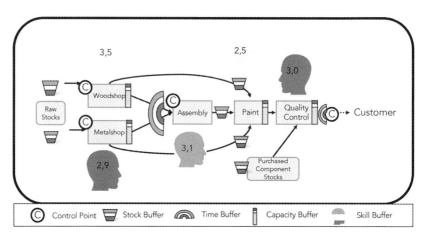

FIGURE A-14 Demand Driven Operating and Skill Model

In this Demand Driven Operating and Skill Model, using the example described in this appendix, the skill priorities to train employees to protect flow are:

1. Red head in Quality control that doesn't yet have a training plan in progress.
2. Red head in Metalshop where there is obviously no trainer available, but an external training is in progress.
3. Yellow head in paint where the number of skill/person is lower than in other departments (that is because the number of people able to paint is small).
4. Yellow head in the Woodshop with a higher skill/person.

5. Green head in assembly is the lowest priority even if the skill/person is low because flow is protected by an expert and one trainer who is actively training new employees.

Adapting the DD Skill Model

The mechanism that allows the complete Operating Model, including skills to adapt to strategic decisions communicated by the business plan, is the DDS&OP process. This bidirectional tactical reconciliation hub defines and communicates the master settings for the DDOM; with the DDSM it can now also identify missing skills and decide the allocation of training resources to better protect and promote flow and to welcome innovation.

Once the managers of the company are clear about the missing skills per department and understand that education is the first step toward protecting and promoting flow and enabling adaptation to the market, the CEO can ask them to show the progress of their training plan during the Adaptive S&OP process. Figure A-15 shows the role of each transverse process of the Competitivity–Competency Plan to protect and promote flows. The Marketing-Sales process as well as the Supply Chain process are the operational processes; the Finance and HR transverse processes are there to support the operational ones; and the Total Quality process role is to formalize all processes.

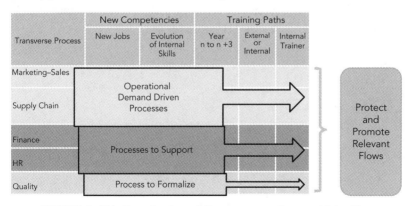

FIGURE A-15   Contribution of Competency–Competitivity Plan to protect and promote flows

In the DDS&OP agenda, evolution of skills needed to protect flow is described by the manager/owner of the operational transverse processes with the metrics of the DDSM the same way they are used for the other buffers of the DDOM. The S&OP committee can decide as a team, with the support of the members from the Finance and HR departments who studied the ROI of the training. They can provide data and explain where to allocate resources for training as a priority. This means the end of the perpetual postponement of training which is experienced by many companies. When the process evolves and integrates new skills, the Total Quality department can validate these new skills when formalizing the new processes. The more internal trainers available, the quicker the company can adapt to innovation, hopefully before their competitors.

Summary

Why do we need to manage skills in a company? Because they can slow down or even stop flow when they are insufficient or missing. This appendix introduces a fourth buffer in the Demand Driven Adaptive Enterprise Model (the skill buffer) and defines the Demand Driven Skill Model (DDSM). The DDSM enables a company to decide where to locate and how to size the skill buffers depending on strategic priorities to better protect and promote flow and for successful innovation. The Demand Driven S&OP process can then configure the Demand Driven Operating Model as well as the Demand Driven Skill Model based on the output of the Adaptive S&OP process.

Adaptive Systemic Thinking

by Chad Smith

In Chapter 2 of this book we asked some critical questions pertaining to the capabilities of organizational personnel:

- Are people in your organization trained to think systemically? Without the capability to think and problem solve systemically, innovation and adaptation will be severely limited.
- Do they have a common systemic language and framework to think and work within? With different vernacular, nomenclature, and modes of operation for similar activity across different areas, it becomes difficult for one area to relate to another—there must be constant translation.
- Do people in your organization understand the connections between departments, functions, resources, and people? Without understanding these connections, personnel cannot understand the total picture of flow. They may take actions that seem good from a local consideration but are devastating to general company flow.
- Are people given enough visibility to understand the connections between departments, resources, and people? Without tools and

processes to ensure visibility, personnel cannot keep evolving their knowledge of the system.

- Are people discouraged from thinking and offering solutions outside of their operating area? If people are discouraged from thinking globally, they will only think about local benefit.
- Do people understand how the different forms of variability affect the enterprise and *flow* through it? Without the ability to understand where to manage variability and what variability to manage, people cannot take the necessary steps to protect flow in the system.

If people do not understand the systemic view or cannot see the system holistically or are actively prevented from acting in the interest of the entire system, an organization is incapable of promoting and protecting flow. Tools and methods to enable people to see a bigger picture and manage connection across the system have been a primary feature of this book. While the Demand Driven Adaptive Enterprise framework has three well-defined components designed to establish and maintain a system based on flow to drive adaptation, there is still the need for systemic and breakthrough thinking to help drive innovation (internally or externally) or even begin the Demand Driven journey (move from Stage 1 to Stage 2).

The Need to Think and Problem Solve Differently

We now turn our attention to a thinking process that adaptive agents can use to drive innovation and adaptation. Innovation and adaptation are driven by identifying and overcoming obstacles and risks to the system's ability to manage, grow, and sustain flow. Many of these obstacles and risks often remain hidden under the surface and obscured from view in the routine "noise" of the system. This appendix focuses on a strategic thinking process designed to reveal and mitigate those obstacles and risks called adaptive systemic thinking.

Most organizations have critical and fundamental problems addressing issues systemically. In most cases, organizational management and other personnel tend to study, construct, and implement solutions in isolation or confined within their specific areas of the system. We know that our organizations and functional areas of responsibility represent dependent systems, yet when it comes to measuring, operating, and problem solving, we continue to divide or segment them into pieces. Under this basic modus operandi it is nearly impossible to maintain coherence to organizational flow, and identify innovation opportunities now and in the future.

Additionally, this type of approach generates a tremendous amount of waste or drag on the organization's inherent potential. Shorter term solutions are typically deployed (sometimes at great expense) to address a specific issue without consideration for the impact on the rest of the function or organization. This type of solution deployment seldom addresses the real issue and often only serves to move the problem to someone else's area of responsibility. Is this done with intent? Of course not! It is extremely difficult to sort through the noise and confusion of the fires to be fought and the intense pressure on management to address specific issues immediately for organizations. Management cannot see what they are doing to themselves; it is the ultimate form of self-imposed variability and waste.

The Adaptive Systemic Thinking Process

This adaptive systemic thinking process was developed over a 20-year time span and has proven itself across a wide variety of environments. Based primarily on the thinking processes of the Theory of Constraints, the process is a streamlined but extremely effective vehicle for an individual to learn a process of strategic thinking that will leave an indelible mark on the way a person looks at problems and solutions within a system.

There are five steps to this process.

- **Step 1: Find a systemic problem.** Individuals and/or teams learn to make the connections across organizational symptoms that are potentially separated by function, location, etc. and identify a systemic core problem that results in the persistence of those symptoms. Many organizations and managers fight the same fires on a frequent basis, waging the same tired battles over and over again. Intuitively, we know that as long as we continue to observe the same types of symptoms, we are not addressing the real underlying core problem. But how do we find the real problem? In an organization or functional area that continues to struggle with the same types of challenges or symptoms, there are often underlying and unresolved dilemmas or conflicts that tug people in at least two directions simultaneously. These dilemmas result in a constant and unsatisfactory set of compromises at all levels producing consistently poor or underwhelming results. Frustration is a daily occurrence and the best people in the company may leave rather than continue to deal with this Sisyphean challenge.

- **Step 2: Spark a breakthrough idea.** Individuals and/or teams learn to identify a specific idea that directly challenges the core problem. A breakthrough idea allows an organization or an individual to break the underlying dilemmas by attacking and invalidating the assumptions that hold them together. This yields a place to start to build a future state, one that does not include the current symptoms. It is a solution that is elegant, meaning it is both simple and extremely significant. The simplicity comes from the fact that we are attacking the symptoms at fewer points (through the underlying dilemmas) and the significance is defined by the knowledge of how the symptoms are connected to and dependent upon the dilemmas.

- **Step 3: Construct a full and immune solution.** Individuals and/or teams learn to construct a complete solution driving to the required results while ensuring there are no disastrous side effects. A breakthrough idea is just a starting point. Other solution elements must be identified and incorporated to complete the solution and mitigate any potential risks or side-effects.

- **Step 4: Plan to address obstacles.** Individuals and/or teams learn to develop an implementation path not based on arbitrary milestones but on the identified significant and specific obstacles to be encountered in the environment regarding the implementation of the defined solution. This path should be carefully sequenced, identifying immediate primary measurable objectives to achieve.

- **Step 5: Implement and manage the change.** Individuals and/or teams learn a basic communication tool set to use with people not involved in the solution development to facilitate and adapt the plan throughout the implementation project or process.

A cross functional team that learns this process together grows their effectiveness together. They learn a new way of thinking, communicating, and problem solving collectively. They can better see connections across the company and begin to break down silo thinking to develop and apply solutions systemically.

Applying the Process—The TreeCo Example

To demonstrate this process, we will use an example based on an actual analysis done by a team of four individuals from a prominent wood products company in the United States. We will call this company TreeCo. TreeCo has a variety of manufacturing facilities that make a wide variety of products including plywood, dimensional lumber, particle board, engineered wood products, and oriented strand board. Additionally, they own nearly one million forested acres in the western United States in relative proximity to their manufacturing facilities.

Note: The space allotted to this appendix does not allow every step to be explained in its entirety. Instead we will focus on the first two steps. For those readers wishing to see the whole story, it will be available at www.demanddriveninstitute.com.

The process must begin with a clearly defined subject matter that has to meet the following criteria:

- The person or team performing the analysis must have intuition and experience in the subject matter area. Without a relatively substantial knowledge base around the topic, the process will bog down and simply produce more questions than answers. That, of course, can sometimes be helpful from a longer-term perspective but it means that the process will be stalled or incomplete until those questions are answered.

- The person or team performing the analysis must have an ability to affect change within the topical area. This is not to say that the person or team has to have absolute authority, but they must have a reasonable degree of influence within the area.
- The person or team performing the analysis must have some amount of energy and passion for the topic. Without that energy and passion, the process will simply stall or be applied in a half-hearted fashion. Simply put, if a person or team does not have enough energy and/or passion about a certain topic then they likely will not be an effective adaptive agent(s) on the topic.

The TreeCo team consisted of four individuals:

- The Plant Manager of TreeCo's largest manufacturing facility (one of the largest plywood mills in the world)
- The Corporate Controller for all of TreeCo's manufacturing assets
- The Director of Quality for all of TreeCo's manufacturing assets
- The Director of Human Resources

The subject matter is most often verbalized as a question or a stated objective. The TreeCo team begins with the following defined subject matter:

We will be one of the most financially successful closely held forest product companies in North America with our existing asset base.

An alternative verbalization in the form of a question would be:

How can we become one of the most financially successful closely held forest product companies in North America with our existing asset base?

The verbalization of the subject matter implies that they want to drive growth with what they currently have available as opposed to a growth by acquisition strategy. Given the team's collective knowledge of TreeCo's manufacturing assets, their current capabilities, and their financial conditions, this subject matter meets the necessary requirements to move forward.

Once the subject matter is determined, the next step is to make a list of between five to ten symptoms related to the subject matter area. These symptoms should be verbalized in a way that meets the following criteria:

- Known to exist
- Recurrent and bothersome to the person and/or the organization regarding the subject matter
- Quantifiable and/or observable
- Not speculated effects or speculated causes
- Simple statements; no compound sentences

The TreeCo team has compiled data that tells an alarming story. Profits are abysmal and not shared publicly due to the company being private. Due date performance hovers between 50–60 percent. Inventories balloon and then either rot or are sold off at very low cost. While labor costs per hour have remained largely steady, the cost to produce each unit is growing steadily, producing poor margins and low returns. ROCE was 3 percent last year. Large retail customers are irate and threatening to source elsewhere. Finally, cash flow is extremely hard to manage, causing constant borrowing. The Controller shares with the team that TreeCo is in imminent danger of breaking its banking covenants. The team estimates that TreeCo has spent nearly $11 million and 112,000 employee hours attempting to combat these symptoms over the last four years to no avail. Frustration throughout the enterprise is at an all-time high.

Below is the symptom list chosen by the TreeCo team, all of which are backed up with actual data.

- TreeCo Profits are low
- Poor due date performance

- High veneer inventories
- Manufacturing costs are too high
- Poor ROCE (return on capital employed)
- Poor margins
- Poor customer service
- Cash flow is poor

Now the team must begin to connect the symptoms. Some connections are obvious, others not so obvious. Yet if they are all truly part of the same subject matter they must all connect in some way. The question becomes, how can the team find the common cause or core problem that connects them all as quickly as possible?

The team uses a process of correlation to attempt to quickly find the connection between all these symptoms. This process is rooted in the following logic:

- All symptoms within a given subject matter must be connected in some way.
- Each symptom brings with it a dilemma when it is encountered; dealing with it typically forces a compromise or sacrifice of something valuable to the organization.
- When looking at these individual dilemmas side by side, a pattern tends to emerge revealing a larger underlying systemic dilemma.
- This larger systemic dilemma oscillates between two extremes and ultimately drives the symptoms of the subject matter.
- Massive amounts of time, money and energy are squandered combatting the symptoms if the dilemma goes unaddressed.

The first thing the TreeCo team must do is begin to understand the dilemma each individual symptom brings with it. They pick three symptoms to start: high inventories, poor due date performance, and poor margins.

To define the dilemma involved with each symptom, they must ask a few key questions in a prerequisite sequence. As in many things in life, the power of the question determines the power of the answer. This process helps people surface their intuition or assumptions about the symptom. The key to ultimately resolving the symptoms is often hidden in that understanding. The four sequential questions are:

1. What typical action do you feel often leads to this symptom?
2. What is an opposing alternate action (or inaction)?
3. What is the objective of the action that you feel leads to the symptom? (Why do it? What is that action hoping to accomplish?)
4. What is being jeopardized by the action in question #1?

Additional background questions can help as well for context if the previous questions are insufficient to surface intuition: Does this symptom ever put you into a conflict? If so, describe it. Why is the symptom undesirable? Can you quantify the effects?

Here are TreeCo's answers for each of their selected symptoms.

Symptom: High Veneer Inventories

- What typical action do you feel often leads to this symptom? We keep peeling veneer.
- What is an opposing alternate action (or inaction)? We stop peeling veneer.
- What is the objective of the action that you feel leads to the symptom? (Why do it? What is that action hoping to accomplish?) We must reduce the material cost of our plywood panels.
- What is being jeopardized by the action in question #1? If we keep peeling logs we lose the flexibility to do other things with those logs (sell them or make dimensional lumber).

Background on This Symptom

The peeling lathe is one of the newest pieces of equipment and has quite an appetite. It can peel veneer much faster than TreeCo can press panels. TreeCo currently has $5.7M in veneer inventory; about half is in danger of rotting. TreeCo will have to sell off the veneer shortly at a bargain price or it will be a complete loss.

Symptom: Poor Due Date Performance

- What typical action do you feel often leads to this symptom? We sell logs to other wood products companies.
- What is an opposing alternate action (or inaction)? We keep the logs for internal production.
- What is the objective of the action that you feel leads to the symptom? (Why do it? What is that action hoping to accomplish?) Maximize the cash value of the log.
- What is being jeopardized by the action in question #1? Meet market demand for finished wood products.

Background on This Symptom
TreeCo can often not meet customer demand in a timely fashion because they do not have the right species of log available to the manufacturing facilities when needed even though TreeCo harvests that species from its own land.

Symptom: Poor Margins

- What typical action do you feel often leads to this symptom? We sell excess veneer to other panel producers.
- What is an opposing alternate action (or inaction)? Do not sell veneer.
- What is the objective of the action that you feel leads to the symptom? (Why do it? What is that action hoping to accomplish?) Convert our excess inventory to cash.
- What is being jeopardized by the action in question #1? Protect panel prices in the market.

Background on This Symptom
Currently TreeCo essentially subsidizes its direct competitors by selling off veneer at low prices. That veneer goes into panels that compete directly with TreeCo panels in the general market. There are a number of panel producers now located around TreeCo facilities that are reliant on TreeCo veneer.

Now the TreeCo team must put these answers into a format that will depict the dilemma involved in each symptom as well as provide the ability to compare across dilemmas. This format is simply a diagram of conflict called a "conflict cloud." The answers from each question are plugged into this format as simple concise statements that summarize the answer to the questions. Figure B-1 shows the conflict cloud structure and the placement of each answer.

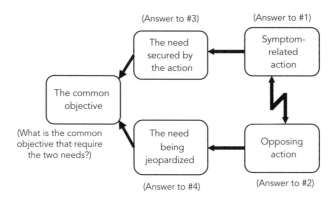

FIGURE B-1 Conflict cloud structure

In the upper right-hand corner of the diagram, the answer to the first question is placed. Immediately to the left is the answer to question three. In the lower right-hand corner is the opposite action or inaction from question two. The answer to question four is then placed immediately to the left of the answer to question two. Finally, at the far left of the diagram is what is called the common objective. This is a stated objective that requires the answers to questions three and four (the identified needs). In other words, the accomplishment of the answers to questions three and four enable a higher objective—they are necessary conditions for that higher objective to happen.

The TreeCo team plugs in their answers from the three symptoms in three separate diagrams. Figure B-2 is the conflict cloud for the symptom "high veneer inventories." Here we see a conflict between continuing to peel veneer for cost reasons as opposed to stopping the peeling of veneer in order to keep logs intact for other potential requirements.

Both requirements have the common objective of using TreeCo assets wisely. TreeCo views both the lathe and the logs as assets and each opposing action is attempting to use each asset wisely.

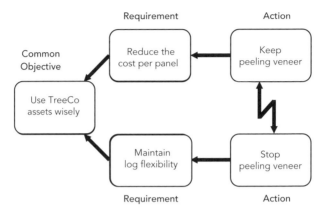

FIGURE B-2 High veneer inventories conflict cloud

While the conflict cloud is built from right to left, it is verified and read in the opposite direction starting with the common objective and the upper leg. Once the upper leg is completed, the reader returns to the common objective and validates the lower leg. The opposing actions are then compared to confirm they are indeed in conflict. Finally, each action is compared with the opposing requirement. This final step is called "squaring the cloud."

TreeCo's initial conflict cloud is read as such:

- "In order to use TreeCo assets wisely, TreeCo must reduce the cost per panel."
- "In order to reduce the cost per panel, TreeCo must keep peeling veneer."
- "In order to use TreeCo assets wisely, TreeCo must maintain log flexibility."
- "In order to maintain log flexibility, TreeCo must stop peeling veneer."
- "Keep peeling veneer is in conflict with stop peeling veneer."

- "Keep peeling veneer <u>jeopardizes</u> maintaini<u>ng</u> log flexibility."
- "Stop peeling veneer <u>jeopardizes</u> reduci<u>ng</u> the cost per panel."

The words and letters that are underlined are the language that is added to make the verbalization flow from the conflict cloud.

Two more TreeCo conflict clouds are constructed and verified using this process. Figure B-3 depicts the other two TreeCo conflict clouds. The reader should practice the verification on their own using the same verbalization from the first conflict cloud.

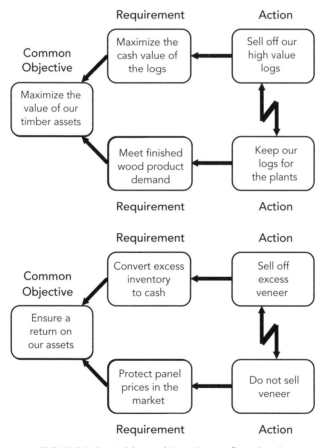

FIGURE B-3 Additional TreeCo conflict clouds

Now a comparison must be made between these conflict clouds. Is there a pattern that is emerging that will help us hypothesize about a larger underlying dilemma at TreeCo? To answer this question, we must compare each component of each conflict cloud to its corresponding component in the other clouds.

First the TreeCo team compares each upper right-hand action with each other: "keep peeling veneer," "sell off our high value logs" and "sell off excess veneer." Two (the first and third) have an obvious connection about peeling and selling veneer. The second action, however, seems disconnected.

The team decides to test whether this situation is a "flip." A flip occurs when one of the conflict clouds is inverted. In other words, the conflict is similar to the other conflict clouds, but it is simply upside down. The opposite (lower) action of that cloud lines up with the upper actions of the other two. When a flip occurs, it is typically a promising sign that there is a chronic underlying conflict.

The team inverts the second cloud. This means that the upper-right actions are now "keep peeling veneer," "keep our logs for the plants," and "sell off excess veneer." Now the TreeCo team sees the connection. All are focused on the manufacturing assets of the company: lathe utilization, the ability to feed manufacturing assets the right log, and the ability to recover cash from running the lathe. They decide to verbalize this general mode of action as "Manage to manufacturing utilization and profit."

Now the team attempts to see if they can characterize the opposing mode of action. To do that they must compare the lower right-hand opposing action of each cloud (while maintaining the second cloud's inversion). The statements are "stop peeling veneer," "sell off high value logs," and "do not sell veneer." The team immediately makes a connection. While the first mode of action was focused on doing the most with the manufacturing assets, this set of actions is focused on doing the most with the log. They decide on the verbalization "Manage to log utilization and profit."

The next step is to tackle the common objective. The three common objectives are "use TreeCo assets wisely," "maximize the value of our timber assets," and "ensure a return on our assets." The commonality is obvious; all are talking about the use and return of TreeCo assets. The team decides to verbalize the generic common objective as "Maximize TreeCo Return on Capital Employed (ROCE)."

Now that the opposing mode of actions and common objective have been verbalized, the TreeCo team turns their attention to the requirements of each conflict cloud (while maintaining the inversion of the second cloud). The team starts with the set of upper leg requirements they verbalized as "reduce the cost per panel," "meet finished wood product demand," and "convert excess inventory to cash." All three seem to have a connection involving the use of TreeCo's manufacturing assets. They decide to go with the verbalization "Drive utilization of manufacturing assets."

Next the team focuses on the lower requirements of the three conflict clouds (still maintaining the inversion). These requirements are "maintain log flexibility," "maximize the cash value of the logs," and "protect panel prices in the market." In this case all statements seem to emphasize getting the most return out of the primary material in wood products: a log. They settle on the statement "Best utilize timber/logs assets." Figure B-4 is the verbalized underlying conflict cloud.

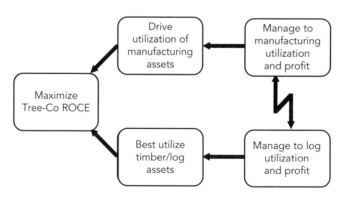

FIGURE B-4   The underlying conflict cloud

After checking the cloud, the TreeCo team is satisfied that the identified conflict does make sense given their experience. At this point, however, the most they have done is to connect three of the original eight symptoms. At best this is a hypothesis about the intersection point of all eight symptoms. But how to test and validate that hypothesis?

The TreeCo team's first step is to attempt to find a sequence in the original list of eight symptoms. Do some symptoms lead to others? Figure B-5 is the symptom map from the TreeCo analysis.

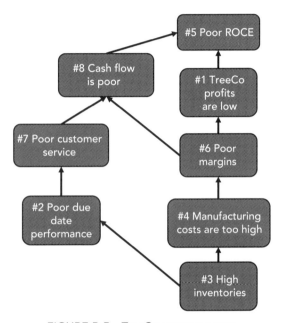

FIGURE B-5   TreeCo symptom map

The team must now turn their attention to connecting this verbalized underlying dilemma to this symptom map. This must be done in a format called a current reality tree (CRT) using sound cause and effect logic that will clearly show how this underlying dilemma results in the previously identified symptoms. A good cause and effect structure tells a compelling story and paints an undeniable picture. Most people can learn how to build solid cause and effect connections through practice utilizing a few basic rules.

The symptom map allows the team to focus its attention on connecting the underlying dilemma to some of the lower or more primary symptoms first. An immediate connection jumps out with the symptom "high inventories" and the verbalized mode of operation "manage to manufacturing utilization and profit." If TreeCo tries to manage to manufacturing utilization and profit, then TreeCo has high inventories. Why? The team immediately determines that the answer is because "TreeCo keeps the lathe peeling all the time."

That makes sense to the team but when asked why TreeCo would keep the lathe peeling all the time the answer was easy: because they are trying to maximize its utilization. This means that they have identified that the most immediate effect of TreeCo managing to manufacturing utilization and profit is not high inventories; it is the behavior or action of keeping the lathe peeling all the time.

The team now attempts to clarify this first connection. If TreeCo tries to manage to manufacturing utilization and profit, then TreeCo keeps the lathe peeling all the time. Why? The team realizes that the answer is that manufacturing up-time is tied directly to utilization and profit calculation. Now we can test this logical link with the following statement. If TreeCo tries to manage to manufacturing utilization and profit *and* manufacturing up-time is tied directly to utilization and profit calculation, *then* TreeCo keeps the lathe peeling all the time. Figure B-6 depicts this first connection on the TreeCo CRT. The arrows move up, connecting cause to effect. The ellipse ties two causes together, making the effect dependent on both causes being present. The ellipse simply means "and."

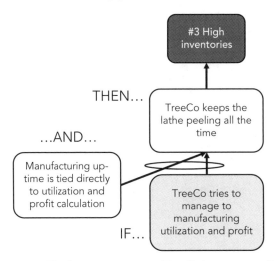

FIGURE B-6   The first connection in TreeCo's current reality tree

Now the team continues to build the current reality tree using the symptom map as a guide and the basic rules of logic. They are eventually able to connect both sides of the dilemma and all eight symptoms. Figure B-7 is the entire TreeCo current reality tree.

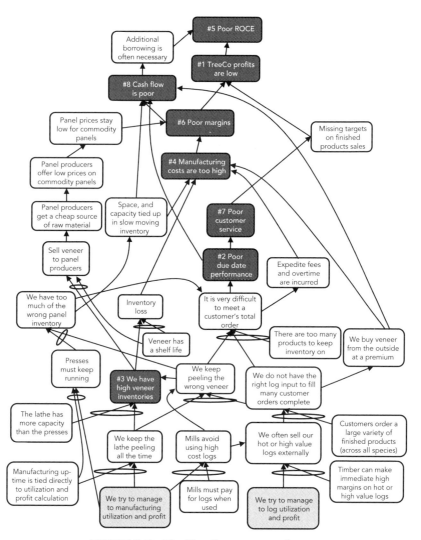

FIGURE B-7   The TreeCo current reality tree

The manufacturing assets and the people that control them are focused on making those assets (and the company) profitable by driving utilization of said assets. At face value this seems to make a lot of sense. However, this directly results in at least three major outcomes. First, they keep the lathe peeling all the time. Second, they keep the plywood presses running as much as possible. Both effects are driven by the fact that cost per unit is reduced by the volume running through those resources. The lathe, however, has more capacity than the presses so the result is a large amount of veneer that stacks up. Perhaps this would not be so terrible if it was the right veneer, but unfortunately that is not the case. It is TreeCo policy that the manufacturing side of the company must immediately pay the timber side of the company when a log is consumed. This typically results in manufacturing assets shying away from using high priced logs (the ones the market currently wants) especially when the plants are near the end of a measurement period.

If the lathe keeps peeling veneer but only peels the "wrong" logs (the ones the market is not really demanding), then TreeCo has significant amounts of the wrong veneer. You can't make the right panels with the wrong veneer, so the wrong panels are produced. This means that the stock of unwanted finished panels raises.

But the presses are slower than the lathe, so the wrong veneer stacks up. Veneer has a shelf life—it will mold and rot if left sitting too long, especially in the damp climate of the Pacific Northwest. This happens with regularity and results in significant inventory loss. When the veneer is reaching its life limit, TreeCo manufacturing attempts to recover as much as they can out of the veneer. They sell the veneer to local panel producers that only have presses (no lathes to peel their own veneer). It turns out that there are several producers in the area that rely on TreeCo overproducing veneer. This is where things get very ugly for TreeCo.

Every piece of veneer they sell to a panel producer results in a panel being pressed that directly competes against TreeCo commodity panels. The competitors get a cheap source of supply and can offer the market lower prices for those commodity panels. This depresses the commodity panel pricing in the market and boomerangs back to hit TreeCo very hard financially.

Now, let's focus on the other side of the dilemma, managing to log utilization and profit. The wood products industry has always understood one thing—the real resource is the log. Getting the most out of the log is the ultimate determinant of sustained financial success. As such, the people in charge of the TreeCo timber assets are always looking to get the most money out of a log. When the market price is high for a type or species of log, they can generate immediate cash from selling those logs to other wood products companies. Why wouldn't they if TreeCo manufacturing is going to delay using them (risking spoilage)?

The ultimate effect of this, however, is that TreeCo does not have the right logs to fulfill all its customer orders. Most large customers order a variety of products (plywood, dimensional lumber, particle board, etc.) together. But TreeCo has too much of the wrong inventory and not enough of the right to meet those orders. This causes delays in shipping and frequent purchases of higher priced materials to make the right things in expedited fashion.

The team has validated the underlying dilemma hypothesis. For the TreeCo team, the full picture of the current reality tree is a staggering revelation. It explains why the company has struggled, continues to struggle, and how off the mark previous improvement efforts have been. Without addressing the underlying dilemma, there is simply no way out. The conclusion is that it is not manufacturing or timber that is *the* cause; it is the way the company has directed these two assets to behave. But what to do about it? How do they escape this dilemma?

The key to unlocking any dilemma is to understand the assumptions and/or conditions that keep the dilemma together. Did TreeCo set up the current mode of operation with the intent to produce this dire situation? Of course not! The system was set up to produce just the opposite. The primary assumption was that this management structure would produce the best overall return by promoting healthy competition for the most critical resource—a log. What the current reality shows, however, is that this management structure is promoting very unhealthy competition that is destroying financial return.

Specifically, how is it that managing to manufacturing profit and utilization comes into direct conflict or competition with managing to log utilization and profit? The TreeCo team turns its attention back to the conflict cloud diagram and attempts to answer this very question. They start by trying to understand the specific conditions that really put the current management of manufacturing and timber into direct competition. They use the verbalization "managing manufacturing utilization and profits is in direct conflict with managing log utilization and profit when. . . ." This verbalization seeks to surface exactly which conditions produce the direct competition. The answers are displayed in Figure B-8.

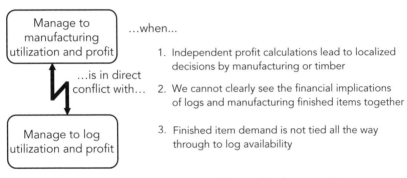

FIGURE B-8 Conditions producing the direct conflict

The team identifies three things that directly lead to the conflict between manufacturing and timber assets. Figure B-8 shows the question and those three answers. Managing the assets independently has led to localized decisions that have seriously jeopardized performance (and systemic coherence). Furthermore, that independence has obscured what is truly the most relevant piece of information—what the assets can really do together.

The TreeCo example clearly depicts a company in Stage 1 of the DDAE development path described in Chapter 8. TreeCo is struggling mightily with an emphasis on unit cost in its manufacturing assets while still maintaining a requirement for better flow; it is a battle that is killing their bottom line and even jeopardizing the company's very existence. Now the door to a solution, however, has been cracked.

Space prevents telling the rest of the story in this appendix. For those readers interested in the rest of the process used by the TreeCo team, it is available online on the "Adaptive Systemic Thinking" page at the Demand Driven Institute's website (www.demanddriveninstitute.com). How will they define a way to move to Stage 2?

Notes

Chapter 1

1. https://www.goodreads.com/author/show/230707.Leon_C
   _Megginson.
2. https://en.wikipedia.org/wiki/Dodo.
3. Martin Reeves, Simon Levin, and Daichi Ueda, "The Biology of
   Corporate Survival," *Harvard Business Review*, January–February
   2016, p. 1.
4. This is an updated version of a strategic target chart that first
   appeared in *Demand Driven Performance—Using Smart Metrics*,
   Smith and Smith, (McGraw-Hill, 2014), p. 206.
5. https://my.clevelandclinic.org/health/articles/10881-vital-signs.
6. Debra Smith and Chad Smith, "*Demand Driven Performance—
   Using Smart Metrics*," McGraw-Hill, 2014, p. 189.
7. Ibid., p. 196.
8. Ibid., p. 197.
9. Ibid., pp. 200–201.
10. Reeves, Martin and Ueda, p. 6.
11. Smith and Smith, p. 154.
12. http://www.apics.org/about/overview/apics-news-detail/2018/
    01/31/study-reveals-need-for-more-accurate-costing-information
    -in-supply-chain.
13. Ibid., p. 14.
14. Ibid., p. 72.

15. Ibid.
16. Ibid.
17. APICS Dictionary.

Chapter 3

1. The logos for the Demand Driven Operating Model, Demand Driven S&OP, and Adaptive S&OP are trademarks of the Demand Driven Institute.

Chapter 6

1. Ling and Coldrick, *Breakthrough S&OP*, 1994.

Chapter 8

1. https://www.merriam-webster.com/dictionary/transparent.
2. Definition of interpretation at http://www.dictionary.com/browse/interpret.
3. https://en.oxforddictionaries.com/definition/intuitive.
4. http://www.dictionary.com/browse/consistent.
5. http://www.dictionary.com/browse/sustainable.

Appendix A

1. The Peter Principle is a concept in management developed by Laurence J. Peter, that observes that people in a hierarchy tend to rise to their "level of incompetence."

Index

GAAP (Generally Accepted
Accounting Principles),
21–25, 27, 142
Gartner, 50
General Motors, 11
Generally Accepted Accounting
Principles (GAAP), 21–25,
27, 142
generation capability, 151
graphs, 150
green zone centers, 5–6
green zones, 5–6, 64, 65, 71, 95

information. *See also* data
"bullwhip effect" on, 29–30
considerations, 124
distortions to, 30
flow of, 12–13
relevance of. *See* relevant
information
types of, 150
information process map, 144
innovation, 7, 122, 134, 151, 162
integrated reconciliation, 143–146
intelligent report, 135
inventory
considerations, 35, 36, 54, 151
decoupling, 62
described, 151
end item, 15
flow of, 15
levels of, 94–95
overview, 15–16
"right-sized," 15–16
stock buffers, 62
inventory positioning, 179
investment, 18

Law of System Variability, 25
lead times, 59, 86, 140, 159
Lean (Ohno), 20, 26
lever point phenomena, 7

Ling, Dick, 119–120
Little's Law, 15

management accounting, 23, 33, 115
management practices
basic necessities, 3–7
conventional approach to, 2,
38–43
corporate survival and, 2–7
fundamentals, 3
management review, 149–150
management team, 123
management view, 128
Manufacturing Resources Planning
(MRP II), 13, 22
Market Driven Innovation
projects, 164–165
marketing view, 128
Master Production Schedule
(MPS), 38–39, 56, 98, 147
master settings, 97, 98–116, 182
Material Requirements Planning.
See MRP
materials, 12, 18, 28, 40, 80
Megginson, Leon, 1
metrics
considerations, 37
flow-based, 37–38, 43, 78–81
strategic, 150–152
tactical, 116–117, 159–160
minimum order quantity (MOQ),
63, 115
MOQ (minimum order quantity),
63, 115
MPS (Master Production
Schedule), 38–39, 56, 98, 147
MRP (Material Requirements
Planning), 13, 38–42, 179–
180
MRP explosion, 77
MRP II (Manufacturing Resources
Planning), 13, 22